My Dear Mother:

CIVIL WAR LETTERS TO DEDHAM FROM THE LATHROP BROTHERS

My Dear Mother: Civil War Letters to Dedham from the Lathrop Brothers

Text Transcribed by Stuart R. Christie

Unless otherwise noted, all illustrations, photographs, etc. are courtesy of the Dedham Historical Society & Museum.
Design/Art Direction: Lisa Breslow Thompson, LisaThompsonGraphicDesign.com

Library of Congress Control Number: 2020919035
ISBN: 9781941573242

Published by Damianos Publishing
Saxonville Mills
2 Central Street, Studio 152
Framingham, MA 01701
www.DamianosPublishing.com

Produced through Silver Street Media by Bridgeport National Bindery, Agawam, MA USA
First printed 2020

CONTENTS

FOREWORD

The Dedham Historical Society was founded in 1859 to "perpetuate the history of New England and especially the town of Dedham." The Dedham Historical Society & Museum maintains an archive of national significance that includes letters, family papers and institutional documents, photographs, newspapers, and maps, some of which date to the earliest years of European settlement in the area. Scholars, researchers, and enthusiasts use this rich collection of archival materials to illuminate elements of our collective past. The publication of this book is the result of one such endeavor.

The Civil War letters of the three Lathrop brothers provide vivid first-person accounts of the war-time experiences of Dedham's Union recruits, men who fought in the most momentous conflict our nation has ever experienced. The letters are unique primary source materials – personal, unfiltered, and raw.

The publication of the Lathrop brothers' letters originated under the leadership of then Executive Director Vicky Kruckeberg. Vicky worked with Board member Stuart Christie, who spearheaded the project, to accurately transcribe the documents. Speaking for the Board of the DHSM, I am deeply grateful for this tedious work and for the persistence they showed in bringing these letters to a greater public audience. Special thanks are extended to Stuart and Vicky.

Stuart and Vicky were assisted by DHSM Archivists Sandra Waxman and Rebecca Carpenter. Additional assistance was provided by Susan Kuecker and Donna Russo. This dedicated team deserves our sincere gratitude for assembling this remarkable compilation of letters.

Thanks are also due to the Committee that brought the letters to publication: Board members Heidi Fieldston, Steve Brayton, Judy Neiswander, Karen Durham, and Bill Flanagan along with Executive Director Johanna McBrien.

The Board also acknowledges and thanks the family of Joseph Pagliuca for their Memorial Gift that supported this publication and Board Member Bill Flanagan for his generous contribution to this project.

Andrea Gilmore, Board Chair

ACKNOWLEDGEMENTS

I want to thank Vicky L. Kruckeberg, former Executive Director of the Dedham Historical Society & Museum, who was the inspiration behind the project to transcribe and publish these letters from the Lathrop brothers. It was Vicky who saw the value of the letters as a record of the contributions of the men of Dedham to Union victories in multiple theaters of the Civil War. Throughout the project Vicky actively participated in proofreading the transcribed letters, as well as helping to decipher some of the more perplexing words and phrases in the letters.

Upon her retirement, Vicky passed the baton to current Executive Director Johanna McBrien, who provided help and encouragement to the project as it moved forward.

During the nearly three years that it took to transcribe and compile these letters, I had invaluable assistance from two DHSM archivists, Sandra Waxman and Rebecca Carpenter. Sandra provided her guidance in standardizing the notations and in the process introduced me to the most interesting magnifying glass I've ever encountered. Rebecca helped with the photographs of the three brothers as well as the maps they drew in their letters.

I also want to acknowledge the contributions of Susan Kuecker and Donna Russo. Susan invited my participation at the outset, and both Susan and Donna provided references and resource materials on the military history.

Finally, I want to acknowledge the support of the Board of the Dedham Historical Society & Museum. This project would never have been completed without it.

Stuart Christie

INTRODUCTION TO THE LATHROP LETTERS

The three Lathrop brothers of Dedham, Massachusetts — John, Julius, and Joseph Lathrop — were descendants of a Puritan minister, the Reverend John Lathrop (1584-1653), who served in Scituate (1634) and Barnstable, Massachusetts (1639). Their great-grandfather, also John Lathrop, was minister of the Second Church of Boston before and during the Revolutionary War. Their father John Lathrop was a minister and served as a Navy chaplain.

In 1928, the Dedham Historical Society received nearly 100 letters that the three brothers sent home during the Civil War to their mother, Marie M. Lathrop, and their sisters Fanny, Jennie, and Mollie. Donated by a nephew, Julius Wakefield, the letters have been well preserved over more than 150 years.

The Lathrop letters span the period from December 1861 to September 1865, when John writes of his post-war visit to the battlefield at Antietam. Along with that bloody battle, the letters cover the battles at South Mountain, Fredericksburg, New Bern, and Port Hudson.

John Lathrop, oldest of the three, was born in Boston in February 1835. He graduated from Burlington College in New Jersey and Harvard Law School. In 1862, he gave up his law practice to enlist in the Dedham Company of the 35th Infantry Regiment of the Massachusetts Volunteers. He saw action as a company commander in the battles of South Mountain, Antietam, and Fredericksburg. He was discharged in November 1863 after experiencing several medical issues. He returned to his legal practice and later served as an Associate Justice on the Massachusetts Supreme Judicial Court. John Lathrop died in August 1910 and was buried in Brookdale Cemetery in Dedham.

Julius was born in May 1840 in Bordentown, New Jersey. The family later returned to Dedham, where Julius graduated from high school and began work as a clerk. He enlisted in the 24th Infantry regiment of the Massachusetts Volunteers, where he saw action under General Ambrose Burnside in North Carolina. In April 1862, he was promoted to lieutenant, and later captain, in the 38th Regiment and served under General Nathaniel Banks in the Army of the Gulf. He commanded Company I, which saw action in the Siege of Port Hudson as well as the ill-fated Red River Campaign. During the Battle of Cane River in April 1864, he was shot while leading his company. Julius Lathrop died three days later. His body was returned to Dedham, and he was buried in the Village Cemetery, where he shares a headstone with his parents. A stained-glass window in his memory was installed in St. Paul's Episcopal Church in Dedham.

Joseph, the youngest brother, was born in Bordentown in October 1846. He enlisted in the 43rd Infantry Regiment of the Massachusetts Volunteers in August 1862, where he rose to the rank of sergeant. He was subsequently commissioned 2nd lieutenant in the 4th Massachusetts Cavalry Division. Near the close of the war, Joseph was captured during the Battle of High Bridge. Released following the surrender at Appomattox, he mustered out in November 1865. Following the war, he married, moved to Ohio and pursued a career in the insurance industry. He died in February 1907 and was buried in Ohio.

The letters reflect the brothers' strong sense of patriotism and support of the Union war effort throughout their service under Generals McClellan, Burnside and Banks. John's account of the battles at South Mountain, Antietam and Fredericksburg, as well as Julius' experiences in North Carolina and Louisiana, are detailed and accurate. The letters also provide an interesting narrative of life in the various camps to which the Lathrop brothers were assigned.

NOTES REGARDING THE TRANSCRIPTIONS

I have attempted to maintain the originality of these letters to the extent possible, including misspellings and punctuation errors. Several letters contain racially derogatory language. I have maintained this language to retain the historical accuracy of the letters. However, I wish to make it clear that this language does not in any way reflect the views of the Dedham Historical Society & Museum and runs counter to our values.

Considering the volume of written material and the conditions under which these letters were written, they contain surprisingly few strikeouts and inserts. When they do occur, I have attempted to show the changes made.

The penmanship, especially in Julius' letters, is also remarkable. Julius wrote in a beautiful "copper plate" style. His letters frequently opened with this type of calligraphy; although subsequently, whether the result of fatigue or lack of paper, his writing often became cramped and much more difficult to read. One example of Julius getting the most out of the paper available was a letter in which he used a cross-hatch technique, initially writing on a normal horizontal plane and then writing over it on a vertical plane.

The letters were written on a wide variety of paper stock and differing letterheads. The paper was usually a cotton bond, which accounts for the letters surviving in such good condition after more than 150 years. And with relatively little ink fading, the letters continue to be readable.

To show the disposition of troops in battles, John and Julius provided several hand-drawn maps of their unit's location relative to others. The maps are accurate and compare favorably with maps in later historical texts on these battles.

Finally, in the few cases of fragmented letters making it difficult to identify sender and/or recipient, I have provided notations to aid in understanding.

Stuart Christie

Joseph Lathrop

JOSEPH LATHROP LETTER

In camp near Baltimore
July 7th 1863

My dear Mother

On the 1st inst I woke you from Fortress Monroe giving an account of our voyages up to that date. The same evening orders came for the 43rd to embark the next morning at 7 o'clock for Boston. Col Holbrook was at Norfolk with his wife seeing the sights and did not return until noon on the 2nd. About an hour before his arrival in camp an nother order came ordering us to report at Baltimore. At 130 PM we left camp, marched to the wharf and went on board the steamer "Kennebec" – a very slow craft by the way. At 3 o'clock we weighed anchor and at 11 AM on the 3rd inst arrived at the wharf. The troops did not leave the boat until 5 PM. Many of the men were drunk so the regt did not appear very well. But Company B received a great many compliments for our boys knew better than to make beasts of themselves, so we marched steady. The Union Aid Society furnished the whole regiment with a good supper. Nine companies were quartered in the large building belonging to the U.A.S. but Capt Whytal didn't wish to have Company D in with such a noisy crowd so we slept in a freight depot where everything was quiet. At 9 AM on July 4th the reg't marched out to this place where we have
about two miles from the city
remained ever since. Our camp is situated on a hill side in a beautiful grove of walnut trees.^ We have been provided with shelter tents – one to each two men. I am quartered with Sergt Haynes.

For two or three days past our Colonel has been trying to get transportation for us to get home but without success, for the wounded, and rebel prisoners have been brought from Gettysburg to Baltimore by thousands within the past three days and we have been kept in case of any outbreak. An order has just come into camp from Gen Negley for the 43rd reg't to report immediately to Frederick Md. Col Holbrook refused to do this and has been placed under arrest. The regiment was drawn up in line and Lt Col Whiton and Major Lane addressed them, stating the case plainly and concluded by requesting that those who thought they ought to go to Frederick, to skip to the right – 235 men complied. You must remember that Government does not consider our time to be out until July 20th, although many of us have been in service for nearly ten months, therefore until the 20th inst we must obey all orders from superior officers. Those men who refused to go to Frederick, are to have their guns taken away, be sent home and be dishonorably discharged from the service – a degradation which I would never endure nor do I believe you would ever wish to see one of your sons ever leave the service in disgrace. The following in the "roll of honor", as Capt Whytal calls it, Co D – Sergts Taft, Lathrop and Haynes,

Corporals Chickering & Marcy, Privates Nichols, Joe Guild, A.G. Baker, Ball, Talbot and Gulucia. All the officers have volunteered to go. The news from the Army of the Potomac is splendid, & this afternoon we have the official report of the capture of Vicksburg. Never did our cause look brighter, and now is the time to give the rebellion a crushing blow. I wish troops are needed to follow up the flying rebs under Gen Lee. If we can destroy his army Richmond will be ours. Now is the time for every loyal man to do every thing in his power to aid the Government. These men of our regiment who have refused to stand by the colors at this present time will be eternally disgraced. Never will they be able to efface the name of "coward" which will always cling to them. Much as I am disappointed at not being able to see you this week I cannot make up my mind to be discharged from the service in disgrace just to get home a week or ten days sooner, for I do not believe we shall remain at Frederick more than a few days. Just as soon as Lee is driven across the Potomac or captured, we shall go home and if we do not have a stunning reception in Boston and Dedham I am very much mistaken. It is barely possible that we may go to Harpers Ferry, but you had better direct "Via Washington D.C." I do not believe there is any chance of our going into battle, for we shall probably guard prisoners or baggage, wherever we may be. Capt Whytal will write a letter to the Committee on Military Affairs in D – giving a more detailed account of our condition. I write in great haste for I must pack up my things as we except to leave tomorrow morning.

Please tell John that I saw Private Batchelder of his company last night. He has been in hospital ever since the 35th left here, but is now well and anxious to rejoin his reg't but cannot get away from Baltimore unless a commissioned officer comes for him.

Also please say to Mrs Baker if you see her at Connecticut corner that her son Addison G – is well and going to Frederick with the rest of us.

Do not worry about [?] Joe J shall take good care of myself. Give my love to all inquiring friends and to those at home. Hoping that you will think I have done rightly in acting as I have.

I remain
Ever your loving son
Joe

John Lathrop

JOHN LATHROP LETTERS

(Letter fragment)

Head Quarters Picket Reserve
Six Miles from Newbern
Sunday March 30. 1862

Yesterday our regiment started off from camp to go out on picket duty and the march out to this place wasn't very pleasant as it was very hot, and being loaded with overcoats, blankets, haversacks, equipment and rifle, all were rather used up by the time we arrived here, but a good night's rest made me alright again. The outer pickets are stations about two miles further on and scarcely a day passes without some skirmish with the rebel cavalry. Tomorrow our company takes its turn for twenty four hours. We are now quartered in negro houses formerly occupied by the slave, and right across the street is the owner's house, where are some two dozen hounds so that last night we were serenaded many times by a full band of these beasts. My dreams were of dogs and rebels, curiously mixed up, and it will be a wonder if this letter isn't in the same manner (speaking of the battle of Newbern) The 24th were the first to cross the open space in front of this battery after the 21st had turned the enemies flanks and we were inside before the last of the "rebs" had departed, and received therefore as they retreated the last volley fired when we were in open sights. It passed over our heads putting one hole in our colors. We leaped the ditch, scaled the breastworks and planted our colors the second inside, then pushing on to Fort Thompson, hoisted our flag the first in its walls There we took several secesh flags.

After all I have a mighty good time out here and don't regret enlisting, still many a hard march have I tramped but as yet haven't dropped out but have kept up in my place, although we haven't been on a march once but what dozens have given out. XXXX

Just to think of a soldier sleeping on a feather bed I did that last nights, having grabbed one in a house and transferred it to my quarters. This is the first time I have seen one since coming here.

Monday March 31.
3:45 P.M

I have just been placed on the outer part of the pickets with good men, so that there is no danger of being frightened at nothing. The men we relieved report the woods full of animals at night. Coons, wild cats, foxes & possums abound. The make night hideous, however I don't think they will trouble me much. In the morning I will report my nights adventures. Just before leaving the Head Quarters of the pickets about 2 ½ miles back the news came of the arrival of the 1st, 10th & 18th Mass. Regs.

April 1. >AM.

Yesterday afternoon about 5 o'clock heard several shots. Marched my men down to the reserve & on arriving found that some forty or fifty of the enemy's cavalry had chased two of the R.S. Battery boys (who are doing duty as cavalry) and had killed or taken prisoner one and wounded the other. As they came down, the leader of the rebels striking at the remaining R.S man, one of our company fired shooting the secesh through the heart killing him instantly. The remaining enemy turned and fled. If our man had not fired as he did we should have lost another as the reb would have killed him in a minute I did not sleep and last night as you can imagine, every one being on the lookout. The 1st Lieut was going to put me on that post on the outer wood but the Capt thought best to place me the furthest off in the other wood so I missed my chance for a shot at the enemy.

Newbern April 2

We were relieved yesterday afternoon, by the 27th Mass. & marched down to Head Quarters, got our whiskey & started our tramp down to the city. Met the 17th Mass which had just arrived from Baltimore They report the 18th & 22nd on their way but not yet arrived. I have found out the full particulars of the skirmish of the 31st_ It seems that two battery men came across a rebel who represented that he was tired of serving in their army & wanted to join ours. He also reported that there were no rebels, cavalry or infantry within several miles, and what there were were very poorly armed. Our two cavalry men brought him down to our pickets & on the strength of what they heard, started up some two miles to look around While slowly riding along all of a sudden some seventy rebels started out of a place near the wood & fired several shots dropping one from his horse. The other started on the back track the rebels chasing & firing. Only one rebel could keep up near so the others cheered him on, following after us fast as possible. The leading rebel was splendidly mounted and soon got near enough to give the battery boy a fearful cut on the head. While trying to kill his foe he came on by our pickets. Reynolds and the others instead of retreating when they saw the rebels coming down, cocked their rifles & when within twenty and just as the secesher had raised his sword to cut the R. S man down, Reynolds fired blowing the reb from his horse as if he had been hit by a cannon ball He never moved, I don't believe he ever saw the man who killed him, At the same time three of the others of our pickets fired on the main body of the enemy wounding one so that it took two men to hold him on his horse. The rebs retreated behind the bend in the road, while we formed expecting every instant a charge would be made We were all ready. I had my revolver all ready to use as soon as I had fired my rifle and if we couldn't have dropped a few it would have been strange. Word being sent three companies soon came up & marched in but saw nobody. I went up and looked at the dead rebel. He was a splendidly formed fellow over six feet tall & stout in proportion. Although dressed as a private there is no doubt but that he was an officer. He was armed to the teeth with two six shooters, two single barreled horse pistols, a Harpers Ferry rifle and a splendid sabre. I have a button taken from his coat, his cap box and a letter from his sister. The man is Andrew J. Reid.

(Picket duty.)

Imagine yourself sitting down by the side of the road in the thick woods with several other fellows near by, not daring to speak aloud or make a fire, and even to light a match, and not knowing at what instant a bullet might hit some one of the picket, and you can see how we were placed I sat a short distance from the others, with my thumb on the hammer of my rifle my revolver all ready in my belt, and thus spent a good part of the night in the silence and along. Now and then a coon or a fox near by would startle me for a minute or an owl would hoot over my head, but after all none of us felt the slightest fear I shall get the outer [?part?] if we go up again. (END)

Camp Stanton, Lynnfield
Aug 12, 1862
2.30

Dear Mother

All right in camp. We shall be mustered into the 35th in a day or two. I say the 35th. This is probably the number of the regiment into which we shall go, but it is not quite certain. We may go into the 38th. Adjutant Loring is now acting here for the 38th. He has written to Capt Rodman about Julius and will doubtless receive an answer tonight. I dined luxuriously on boiled rice, nothing else, but this ad libitum. Slept last night on hay in a tent wrapped in a blanket [?Woke?] at five this morning drilled the men from six to seven and again from 9 to 10. Direct to me

Lieut. John Lathrop
Dedham Company
35 Reg.
Camp Stanton
Lynnfield
I do not know when I shall be up, probably by the end of the week. Ask Joe to bring in my thick boots. Those with while lining inside.

Love to all
Your affectionate
Son
John

Somewhere in Virginia
6 miles from Washington
Aug 25, 1862

Dear Mother

Here I am at last sitting in the hot sun without the shelter of a tent or tree on the accessed soil of Virginia; sacred it is sometimes called, but surely in irony, for a more desolate and blighted region I never beheld. But you do not want to read a dissertation on Virginia I presume and you would prefer an account of my journey. Here then it is. We left Boston at five and reached Fall River about ten. Here the Bay State was ready to receive us, but how changed. Every carpet taken up, and all the furniture gone no beds in the berths. In the soldiers turned on the slats of the berths in the main and ladies' cabin. That is those who could get there. The rest lay down where they could on the deck. The officers had state rooms, but there were no blankets and but one sheet on the beds. My blankets were with the baggage, but by the help of Willard's rubber blanket and my overcoat I did very well and passed a very comfortable night. We reached Jersey City at about ten the next morning and there four of the companies remained until half past twelve, ours among them. We went through Bordentown & Burlington, but I did not see any one I knew. We reached Philadelphia at five and were most hospitably received by the "Cooper Shop Volunteer Association." I have often heard before of this charity but never realized its usefulness. Here we had an opportunity to wash and then were fed on the best cold meats, pickled beets, raw sliced tomatoes, bread, tea and coffee. I never enjoyed a meal more. We came near having a little fun while resting here. On the opposite corner was a rum shop into which some of our men had been beguiled. The Colonel ordered the contents to be destroyed and a universal smash was the result. Our men were at this time feeding, and I had walked over to see that no one was interfering with our guns when I heard the smashing. I went to the front of the building and found the colonel standing in the doorway with only his men on guard before him with covered bayonets and a crowd of men trying to get in to stop the work of destruction. There was another officer present with me. We drew our swords and helped keep the mob back until a company which had been sent for arrived. Soon after a policeman told the colonel he had a warrant to arrest him and asked him if he would come before the mayor and the aldermen without resistance. He replied that if the mayor and the aldermen insisted on it he would go before them but he should be under the necessity of taking them to Washington as he should construe it as a case of treason, inasmuch as it would afford aid and comfort to the enemy. As the authorities of Philadelphia were not desirous of going to Washington under escort they let the matter drop. We march a long distance in Philadelphia and received quite an ovation. The people along the route turned out to a man, woman and baby 'Good bye", "God bless you", "Come back safe", were the constant exclamations. Women brought out water and did all they could to make the men comfortable. Greater hospitality I have never experienced. We left Philadelphia at nine thirty Saturday night, were in the cars all night and reached Baltimore early Sunday morning. Then we had another long march through the city, but here was no enthusiasm. Now and then a small flag would be waived but generally the people stared with indifference, and looked as if they would like to repeat the reception of April 1861. We were fed here by an institution similar to that in Philadelphia but not so perfect in its arrangements. At eleven we left Baltimore and arrived in Washington at two. We rested at the depot until five when we took up the lines of march for our camp which we were told was three miles off. Up Pennsylvania Avenue we marched by platoons crossed to Georgetown and then into Virginia. On and on over dirty roads until the sun set and still no end to the journey. Up hill and down until we concluded that Richmond was our destination. At length we stopped having marched over eight miles on the worst possible roads, the dust so thick that the head of the company was scarcely visible when we were marching by the flank. This was rather rough for men not used to marching, and who were already pretty much used up. Then we turned off into a field and camped for the night without tents. I spread my rubber blanket on the ground, rolled myself up in my woolen blankets and turned in with the sky for a roof. You may be assured I slept soundly. Our tents are not yet up but will be I suppose by night, and then we shall be more comfortable. I saw Alleyne Adams in Washington and asked him to send word that I had reached there safely. I should have told you that although this place is only six miles from Washington yet we were marched over wrong roads so that we went fully eight miles if not more. I am now very well and all right except that I should like to sleep for twenty four hours on a stretch. Love to all at home

Your affectionate son
John

Direct to me
Lieut. John Lathrop
Co. I 35 Reg. Mass. Vol.
Washington
D.C.

By directing to Washington the letter will reach me wherever I may be.

Yours _ John

Camp Whipple, Arlington
Heights, Aug. 31. 1862

My Dear Aunt Jane

A thousand and one things have prevented my writing to you before. I had a pleasant journey on and arrived in Baltimore a week ago this morning. We had a long march through the city, took the cars for Washington and arrived in the latter city at two in the afternoon. Here we remained until five and then took up the line of march for our camp in Virginia. We went through Georgetown; over the aquaduct bridge to Virginia and on over dust roads until we had march eight or nine miles. This was pretty rough for men who had had a long and tiresome journey in the cars. We were all pretty much used up and about ten o'clock we turned in on this dusty soil of Virginia with the sky for a roof. The next day we got our tents and by night we were more comfortable. This camp was called Casey. We remained here for a few days and then moved to our present positions. We are now about four and a half miles from Washington on a hill which is called Prospect Hill. We have a good view of Washington and of many of the forts around. You have doubtless seen in the papers that a great battle has been going on. We could tell here but little about then, as we were as much in the dark as the outside world. On Friday we were called up at half past three and remained under arms until sunrise, but no attack was made. Later in the day our company was ordered to this end of Long Bridge to do guard duty for twenty four hours. This was not bad [?sport?] as we had a chance to talk with many persons from the army. There we heard that the rebels had captured Manassas, but were hemmed in by Pope and Burnside. McClellan was then at Alexandria. All the afternoon immense droves of cattle were going over the bridge back to Washington. Late in the afternoon a long train of ambulances came from Washington, and towards night bearers of dispatches came along from Pope and McClellan. We had orders not to allow a soldier to pass unless he has a pass signed by a Brigadier General. Toward night we stopped the 12th Pennsylvania cavalry which had run away from Manassas. They told all sorts of stories about the loss they had sustained, and the usual story which those who ran away tell, of defeat, and destruction. The next day, yesterday, we were relieved and got back to camp about two o'clock in the afternoon. A battle was going on all day yesterday. We could hear the report of the guns quite plainly. The firing was almost continuous. This afternoon I have learned that we have conquered Jackson and taken him and his army prisoners. [?Thus?] this is I cannot say but there has been no fighting today at least I have not heard the cannon. Last night we were again called up and sixty rounds of cartridges distributed to every man, and all ordered to sleep ready to fight at a moments notice. I am very well and enjoy the life as well as I expected. When you write direct to Lieut John Lathrop

35 Reg. Mass. Vol.
Washington D.C.

Camp Whipple Va
Sept. 4. 1862
3.30 P.M

Dear Mother

Your letter has just been received and if you knew how delighted I was to see it you would not have waited a moment in writing to me after my letter arrived. Our letters reach here in the afternoon. The Orderly of each company goes to head quarters for them and then distributes them, first calling at the officers tents. Now not to have a letter in bad enough, but to have a man come to one's tent every afternoon with a pile of fifty to a hundred letters and run them over and then inform you, while you are watching the motion of his hands with anxious eyes, and trusting that the last at least may be for you, that "there is none today, sir," is aggravating beyond description. I don't understand why you don't hear from Jule, or rather see him. Ask Mr Wakefield to go to the Adjutant General's office at the State House and see why he don't come on. You do not seem to be aware of the fact that your humble servant has been promoted but so it is. My commission as captain arrived today and I am now in command of the company. My men are off today on picket duty so there is nothing to do but new tactics and sleep. To have therefore a commission arrive and a letter from you the same day is gilding refined gold and painting the lily. I wrote to Joe this morning telling him all the news. I did not send his letter however as I intended to send it by a lieutenant who expected to go in. He did not go and the letter will go with this. There has been no firing for the last two days and the impression here is that Jackson has gone up the river with the intention of crossing into Maryland. He has left a small force in our front to keep us in check and to prevent us from pursuing him. Immense bodies of troops have been sent up the river and there is no danger of his getting across. He is however a bold and skillful general and may do us a good deal of harm before he is used up. There is no danger of his making an advance upon Washington as he is no fool. This army would be terribly cut up by the forts even if we had no troops outside. Every hill for miles around this city is surmounted by a fort well garrisoned and prepared to resist the attack of a most powerful force. In addition to these rifle pits are being dug in every direction and we are all ready for a brush. You speak of the excitement in Boston last Sunday, I am inclined to believe that the nearer one is to the scene of action the less the excitement is. For instance last Friday we were at Long Bridge; rumors had been coming of a battle for two or three days before, on that day we stopped a regiment of cavalry which admitted that it had run away. All that night messingers and bearers of dispatches from McClellan and Pope to Washington and from Head Quarters to these generals passed backwards and forwards. The next day on our return to camp I could distinctly hear the firing & knew from its frequency that a great battle was impending, then on Monday and Tuesday the road was filled with stragglers with accounts of our being defeated, and yet there was but little excitement, not half that which I should have experienced at home. Capt Carroll, I learn from some of the 18th who were here today, was killed and buried in the battlefield. This afternoon there was skirmishing some seven miles from here & for a time the firing was quite heavy. We are in no danger however for there is a force of some forty thousand troops between us & the enemy. We have glorious moonlight nights now, though they are rather cold. After ten o'clock an overcoat is quite comfortable. I should like much to be in Dedham tonight and look in upon you sitting in the parlor reading the paper. If Jule comes home tell him to write to me for it is not worth while for me a send a letter to him at Newbern now, for he would probably not receive it. Mr. Geo. S, Hale a lawyer from Boston is here tonight and I shall probably send this letter by him together with Joe's directed to [?T___?]. I may however send it by the mail. Now don't let a week pass without writing to me. I know that when one goes from home although he is missed at first, yet as everything else goes on as usual his absence is soon forgotten and letter to him grow more and more infrequent, but for all this do write and write often and if there is anything to write about I will tell you of it when I answer your letter; if there is nothing I will write still. If you do not hear from me you may attribute it to the fact that the government has stopped the letter as I believe they sometimes do and as I am informed they have done recently. Give my best love to all

Your affectionate son

John

(over)

P.S.

I met Nelly [?M_tey?] in Boston the day before I left. She asked me to give her one of my photographs. I promised to do so and now enclose it. Please send it to Riverdale for her. Don't change it and send her a worse one.

John

Middletown Md. Sept 14. 1862

Dear Fanny

I write just as we are on the point of marching. My last letter was to Mother from Brookville. We left there the next day, marched all that day, and the next & camped that night near Frederick City on ground from which our advance had driven the enemy an hour before_ All yesterday morning we remained in camp listening to the firing. The enemy were driven before our army_ In the afternoon we started again and marched to this place some ten or fifteen miles from Frederick. I propose putting this in the office here but whether you ever receive it is doubtful as the mails do not yet run. The firing has been quite heavy this morning but has now ceased_ One regiment has not yet been in sight of a fight though within hearing for several days. I made a mistake in respect to our brigade in my letter to Mother_ We are [—at—] the 2nd brigade of the 2nd division of the 9th army corps – Sturgis is in command of the division Reno of the corps – and Ferero or whatever his name is of the brigade. Burnside is over all. We have good generals and are confident of success_ We are to start soon and I must close_ Direct as before We have had no mail since Sunday last.

Your affectionate
Brother John

To Mrs. M. M. Lathrope
Dedham
Mass
Camp near Harpers Ferry
Sept 17. 1862

Dear Mother

I am writing to you in the midst of a severe [—action—]cannonading, but not to anticipate I will resume my narrative from where I left off on Sunday on which day I wrote to Fanny and posted the note [—to—] at Middletown, Md. We left camp at N. on Sunday afternoon about two o'clock & marched three or five miles to battle field. As we approached we met the wounded being carried to the rear in numbers. We formed in line in a cornfield, threw off our blankets and prepared for action. We were first ordered into a wood to disperse the rebs who were concealed there. As we approached some shots were fired and returned but in we went. One of my men Henry Tisdale was here shot in the thigh by a wounded rebel. He was not very badly wounded. The wood was very dense, and it was almost impossible to force our way through. In however we went as well as we could but the men soon got separated & when I got out not more than half my company was with me, the rest being Lieut Hill. The other companies were in the same condition. Officers separated from their men & men from each other. As yet we knew nothing of the plan of battle or what we were expected to do. When we emerged from the woods we were order into a field where the rest of the brigade were drawn up in line of battle by regiments, our regiment being the rear. I had not got my company into line when the firing commenced. It is said that this volley was by the New Hampshire 9th who were drawn up in battle on the other side of the road. I ordered my men to lie down. They did so and returned the fire. Soon the cry was raised that we were firing into our friends. However this may be it was certain that we were fired into first. When the firing was over I attempted to reform my men when I was ordered to take them into the woods behind and form them there. I did & had just got them into line when a murderous volley was poured into us by the rebs. Two of my men were struck. For a short time all was confusion The bullets whistled over our heads, singing like very demons. I will now give you a rough sketch of the battle ground which will enable you to understand our position better.

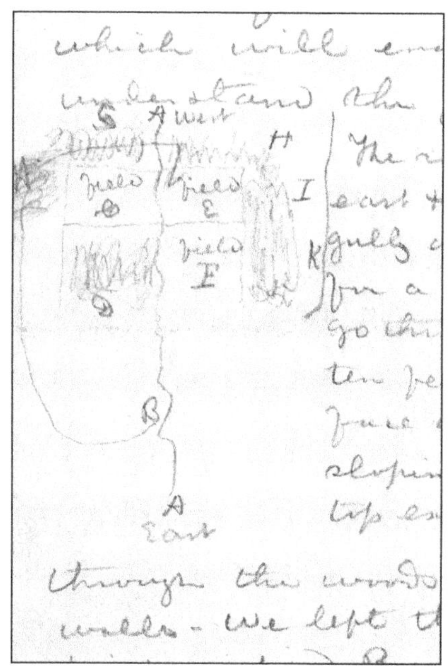

The road A.A. running east & west is a mere gully wide enough for a single [illegible] to go through. It is about ten feet below the surface of the ground with sloping banks. On the top except where it goes through the woods are stone wall. We left the road at the point marked B and the line shows our course through the woods until we emerged into the road. Now to resume. We were soon ordered out of the woods D. To the road to keep the rebs from

F.

flanking us from the field [—E—]. At this time the rebs were in the woods G. & our troops in the

^

field C. It was supposed also that they were in the woods H.I.K. There we lay for over an hour the bullets whistling over our heads. About eight o'clock the firing ceased & we camped for the night in the field (F.) I had then only 14 men of my company with me both of my lieuts. were missing. The other companies were in about the same condition. As soon as the men turned in Capt Oliver & I went down the road to the hospitals to see if we could find any of our men. At the first hospital I found Geo. Whiting having his leg amputated, & Tisdale with a ball in his thigh. At the next Clinton Bagley of South Dedham who was struck in the side by a ball and Lieut Hill who was struck in the leg by a spent ball. He was only lamed. The skin was not broken. He has gone to the rear but will be with us in a few days. Capt Oliver found here one of his lieuts. Williams wounded seriously in the breast. About three o'clock I got back to the field & turned in under a rebel blanket. We gained a great victory. The next morning I took a look at the battle field. Dead rebels were in every direction In the road where we were that night were over fifty. We could not move without stepping on them. The fields and woods around were filled with bodies and knapsacks haversacks, guns & all such things. Such a sight I never beheld. There we remained until afternoon & then the order was forward. We marched until after dark & turned in in a ploughed field. The next morning yesterday, a battery opened on our camp. Several shells [—burst—] struck near us and others went over our heads. Some seven or eight men were killed. One in a regiment in line directly in front of ours had his leg taken off. In the afternoon we started again but only march about a mile when we went to camp. Our batteries are all round us, & the rebels' in front. Last night I witnessed what I have often wished to see an artillery fight by night. The sight was magnificent. I could see the flash of the (Here we were ordered to fall in & our regiment is drawn up on a hill with batteries on each side & the infantry firing a few hundred feet in the valley below. If I get through safe I will continue the letter. If not Good bye. We shall meet again.) Sept 18, AM We had a sever fight yesterday. Our regiment suffered terribly. Nearly all the officers were wounded I escaped without a scratch. Today we shall probably have to stand it again. I can write no more now as we may be ordered in soon.

Sept 18 4 P.M

Dear Mother

We have drawn up in line of battle all day but there has been no firing on either side as yet. Each wants reinforcements. I will therefore resume. I was about to say yesterday when the order was given to fall in that I could see the flash of the gun then that of the explosion of the shell and then the two reports would come along. This was owing to the distance I was from the firing Yesterday I could see the explosion, then hear the sound of the explosion and immediately the rush of the shell over my head. Sometimes latter is heard before the explosion. For an hour yesterday we were drawn up on a hill with a battery on our right Our position was a very commanding one. The battle field was but a short distance below. We could see every shot from our battery strike and also not a few of the rebel battery strike near us_ Then the order came to forward. Down the hill we went & formed on the line of a creek, the rebs being on the opposite side. After this we crossed a bridge which was taken by our side & formed on the brow of the hill. There we lay all of the afternoon with the balls singing and screeching over us until about 5 PM when we were ordered to forward We had first to ascend the hill & then go down at the double quick as far from our house to the bridge. All this time the regiment was exposed to the heavy fire of the rebs. Shot & shell & bullets were pouring into us. At the bottom hill we formed behind a rail fence the rebs being [—in a h—] behind a stone wall some two hundred feet distant We staid here an hour when our ammunition giving out & no more being at hand & no reinforcements sent to our relief we retired to the brow of the hill. Just previous to this I had gone part way up the hill to see Capt King who was acting as major to get permission to send for ammunition, but I could not find him. I then commenced to take the cartidges from the dead & wounded & to tie up the woundsz as well as I could when I saw the regiment leaving. I was at this time near the top of the hill, and had just got the leg of a man tied up when a ball struck him in the other leg & the balls flew around like hail. I was covered with dirt as I stooped over the man I then went to the bottom of the hill to see if my company had left I found it had & returned up the hill. I was the only one on the hill as the time and the evening made me as especial mark. I escaped unhurt. The regiment is fearfully cut up. In the fight on Sunday our Colonel was wounded. Yesterday our Lieut Colonel & nearly all the officers were wounded. The major was not with us. There were but three captains left, Cheever, Andrews & myself. Colonel Fererro this morning asked me to take command of the regiment but as I don't want responsibility I declined & told him I should prefer an older officer. He appointed Capt Wright who has been acting as major of the New York 51st. I was more exposed than [—of—] any officer in the regiment except the lieut colonel & Capt King & my escape is a miracle. The battle of Sunday is called that of the South Mountain. Our loss was as

follows, this is in my company – killed none
Wounded
Lieut Hill struck by a spent ball on leg. Serg. Tisdale flesh wound in thigh Private Geo. Whiting in leg & leg amputated. Clinton Bagley wounded in side. Several were missing and were so reported but all have returned. Henry Ellis among them.
Our loss yesterday was
Killed Patrick Walsh
Wounded Corps Hutton, Davis, Wheeler
Privates Ayer, Cleavleand, Curran, Downs, Alfred Ellis, Holland, Rausch, Joseph Smith of Weston supposed mortally. N.C. Treadwell, J.R. White, Robert White, Hammon.
Missing Ch. Sulkoski.
All the wounded of my company were taken off the field except poor Hatton. He was struck in the back while retreating I went in the night to get his body but our pickets would not let me pass as the field was in possession of the enemy and I might be fired on.

Sept 19

We were reinforced yesterday and our regiment is temporarily relieved. I have not yet been able to ascertain anything about Hatton as the field was yesterday [—in—] neither in our possession or that of the rebs but threatened by both and orders were given that no one should go on. No fight yesterday. I shall write again soon. Ned Davis was wounded in the leg. I feel badly.

Your affectionate son

John

We have had no mail for nearly two weeks

In Camp near Antietam Md.
Sept 22. 1862

Dear Fanny

Your letter of the 10th reached me today. We have had no mail for two weeks except a small one which reached here last night & by which I received nothing but a Dedham Paper. This I was glad enough to get but wanted some letters. However today the chaplain arrived with a big mail by which I received seven letters. My last letter to you & to Mother have kept you informed as to our movements and those of the army so far as I knew anything about them. The day I wrote to Mother we left camp marched over our battle field and towards

side

Harpers Ferry. We are now about 8 miles from the latter place encamped on the [—top—] of a hill with Antietam Creek at the bottom. The place is pleasant enough. The facilities for washing excellent – witness the fact that yesterday I washed my shirt & underclothing and went for an entire day with nothing on but my shoes, coat, pants & vest. full dress. better at least than that southern cavalier which consisted of a hat and a spur. An other objection is that on turning under, I cannot say in for we have no tents, our blankets, we realize the proverb that the way of the backslider is hard, for during the night six feet is a slight descent to make. Last night I slept in front of instead of behind the stack of arms & by bracing my feet against them succeeded in preventing a lapsus. How long we shall remain here or where we shall go next I don't know. There is a rumor today that [—that—] the rebs have come back to Maryland. If so we may have some more fighting to do. If our regiment has the same luck as before you may see me yet at home with my arm in a sling. How interesting it would be. My escape in the last fight is a miracle. Poor Halton was decently buried, and last night I read the burial service to the company in memory of the departed. I feel his loss very much. He march very near me in the ranks being on the right of the company. I am very sorry that Mother let Joe enlist. I would not take him with me because I wanted him to stay at home. Poor Mother, home is now effectually broken up. You must do all you can to cheer her up. I wrote her a full account of our late fights. Major Willard was not in them being at Washington. It is too bad that Jule cant get on. There is no use trying to [—g—] do anything with the war department here. It would take six months to carry it through. The only way is for his colonel to let him go on his own responsibility. This is what our colonel told me a short time before we left Camp Whipple. The thing is done in this way every day. Our mess was given up before we left Camp Whipple & I am now on army rations. We have plenty of coffee, sugar, hard bread, & meat, not always fresh however. These with what we get by foraging do very well. Write soon for letters from home are very precious. Your affectionate brother John

Boston Sept 25th 62

Dear Bro. John.

Your letter to Mother relieved us from great anxiety for you. We saw the lists of the killed & wounded officers from your Regt before it came without your name among those and therefore took comfort but as the battles were so continuous we knew not to what point the accounts brought us. Your escape was wonderful and we thank God for his preservation of you in the midst of such peril. It was a terrible battle. That of Wednesday at Antietam. Also of Sunday at South Mountain. Tho' to us that of Wednesday seems like the old battles of History a hard fought nearly a drawn battle but on the whole a victory for us.

The great disaster of the campaign seems to have been the surrender of Harpers Ferry – but we can judge but poorly at this distance with so few actual facts. On the whole we call the weeks fighting a victory to our arms and disastrous to the rebels, their hopes, and aims. We are distressed at the loss of poor Hatton – the papers say "wounded (since dead)" – I infer from your letter that such must be the case – George was just in having only seen it in the papers – he is anxious that the place of his burial may be marked so possibly they may get his body.

I know [—Mr.—] Pope your 2nd Lieut.s father – he has been in after to learn if I had any tidings from him by you. Pope is a clever man and I hope his son is a faithful and efficient officer. We are all well. Julius returned to his Regt 38th – near Baltimore last Monday having been at home about a week – of this you may have heard before. Joe is in Camp at Readville. Mother and all well – [??] the babies – I sent your money by Adams & Co [?Express?] and they will have it at their office in Washington if you have not got it. I ? written of this before but lest you may not have received it I write it again I send this by so are going directly to your Regt as it is

The citizens of Dedham are full of anxiety to hear from our boys with you – all of a great number have been to learn what you wrote. Your list of casualties from your company was the first received. And the editors of some of the papers came to obtain them.

I must stop as Mr. Pope has come to take this letter to the messenger

Your Bro
T. L. Wakefield

Write —

In camp near Antietam

Sept 27 1862

Dear Sir

I have received your letter of the 25th tonight by the hands of the Bartholomew. If our letters could all come through as speedily I should rejoice. Our mails are very irregular We have had none for several days; and as for females, I have not seen one since I left Brookville that was worth looking at. Oh for a week in Dedham with a party every night. Here we have to be content with balls of a different kind. We left our camp yesterday under the impression that we were going to Harpers Ferry, but after making a circuit of two miles we camped nearby opposite our old place on the opposite side of the creek. It is said that we shall probably remain here two or three weeks. McClellan is to take up his quarters near here & Burnside's head quarters are but a short distance off. I have seen him several times. He is a splendid looking man. The pictures of him are very accurate. The rebels are said to be in force on the other side of the Potomac. The victory at Antietam was a great one and the battle was far from being a [??drawn??] one. It was not renewed on Thursday because our troops were tired out & we waited for reinforcements. It was as if the rebels occupied the stone bridge on the road below our house in Dedham & the land beyond & we should drive them from it, take powder rock, Fullers Hill, & then fight for the field beyond& finally have to retire to the top of Fullers Hill & the next day the rebels after have in the night placed pickets in the field, should withdraw then & retire to Boston. The battle was a fearful one & the loss of life very great. My own escape was wonderful. Lieut Hill has gone to Boston. I asked him to call on Mother. Today I rigged up a shelter with a woolen blanket and shall sleep more comfortably than usual I hope. Last night was very cold. I slept in the open air & this morning my blanket was covered with frost where my breath had gone against it. Edward Hatton's grave is properly marked. There would be no difficulty in procuring his body as soon as cold weather sets in and I dout there would be now. Some one would have to come after it. Mr Bartholomew expects to take back some of the bodies of the Roxbury boys. He could tell you about it. There is a prospect of our being more comfortable soon. We are going to get our trunks from Washington. One trunk will have to do for the officers. Within a few days we have got up a mess. Our principal food is beefsteak, & hard bread first soaked in water & then fried with pork. It is not bad. Tomorrow we indulge in baked beans. Some dessicated vegitables arrived today which we shall partake of soon. Our loss in the late battle was about 48 killed & 140 wounded. We muster now about 500 men all told. A loss of half in five weeks. Such is life. I shall try & send tomorrow with this letter a copy of the first volume of Hardee, rebel edition, which I found on the battle field of South Mountain.

Please give it to Mother to keep for me. Mr B. is going & I must close. Love to all. Write soon
Have an eye to the insolvency case of Gardner A. Fuller

Truly yours
John Lathrop

East
In camp {—West—] of Antietam Creek
Md. Oct. 6. 1862

Dear Mother

Here we are in the same old camp that several of my late letters have been dated from; but the camp has undergone a change. Yesterday the men received shelter tents and the officers wall tents, one for each company. Last week I received on Wednesday your letter of the 17th & 19 and one from Jennie dated the 21st Jennie's I answered the next day and intended to write you at the same time but was prevented by regimental duties. (On Friday the 9th Army Corps was reviewed by President Lincoln, McClellan, Burnside and their staffs together with some members of the Cabinet were present. About ten thousand troops were in the field, our regiment among them. Lincoln looked very care worn. He is homely enough but not so bad looking as I expected. Yesterday was with us a busy day in spite of its being Sunday. In the first place in the morning about nine o'clock an enormous mail arrived. The rush for letters is so great when mail comes that the chaplain is unable to attend the duty alone, and usually one commissioned officer from each company assists. Accordingly, yesterday, ten officers and the chaplain set to work. And we were an hour at least in sorting the letters. My share was a liberal one, fourteen letters and three papers. Many of the former were however on business matters, inquiries respecting the dead and wounded, three being from Mr Hatton. I was the rest of the morning reading the letters and papers and about answering the most pressing when it was announced that the tents, overcoats, shoes and other articles of clothing for the men had arrived and were to be distributed to the company officers at once. This took up all the afternoon. By dark my tent was up and half full of shelter tents and clothes. In the evening I was busy answering business letters until half past eleven. This morning I have been hard at work distributing the things to the men, and now, 2 P.M. I take the first opportunity of
enclosed
answering your letters. First please excuse the envelope this will be [—excused—] in but I am all out of plain ones and these are all I can get from the sutler. Please don't publish any more of my letters or extracts from these especially after a battle as they are apt to be full of mistakes being written in a great hurry merely to let you know that I am safe, and oftentimes while under fire as was the case with the last. The letter as printed stated that the distance that the rebels when behind the stone wall were two hundred feet from us. If I wrote it so, it was a slip of the pen. I should have said yards. Lists of the killed and wounded are always at the disposal of the public, but nothing more. In fact there is an Army Regulation prohibiting officers from being correspondents of newspapers. I received yesterday a letter from Aunt Jane, Fanny, Mary, Jule and Mr Babcock. These I shall answer soon. I shall write to Fanny as soon as we move which it

is said will be tomorrow. We are going I believe to Pleasant Valley near Harpers Ferry. I do not think there will be an advance movement made for some weeks; probably not until more of the nine months men come in. It is a shame that Joe is only a 4th serg. He ought not to have gone into the company without the promise of at least a second sergeant's berth. Now there is no chance of promotion unless indeed his regiment meets with our fate. I am now quite well. Both William and Loring say I am looking ten per cent. Better than when in town. Kiss the dear children for me and tell them that Uncle John will never forget them. Love to all. Take good care of yourself and do not worry on my account. I shall send you some money for yourself when payday or rather when paymaster comes, for the two are not co-temporaneous. At present I am about destitute not having received yet the money Mr. Wakefield sent though I hope to have it this week. Direct as before.

Your affectionate son

John

I expect my trunk this week and when it comes I shall have plenty of envelopes. The package sent in John Cobb's box has not yet arrived.

JL.

Mrs. M. M. Lathrop ,

I sent Edward Hatton's watch home by Major Fay of Chelsea. He will leave it with Mr Wakefield. I had to pay $1.50 to the man to whom Edward gave it to fix.

In Camp Pleasant Valley, Md.
Distant from Harpers Ferry
according to statement
of inhabitants from 2 ½ to
5 miles

Oct. 9. 1862

Dear Fanny

I wrote to Mother the 1st of the week and promised that you should receive the next letter and here it is, written out-doors with spiders and other creeping things trying their powers of locomotion over the paper, and what is worse over me. Such is life. We left our old camp last Tuesday and had a very hard march over Elk mountain to this place. The road was very rough. Every few minutes there was stop ***[Note: The first page ends here and there is an interjection across the top of the next two pages as follows:]***

I should have liked very much to have seen Kate Lincoln. I have not seen her since she returned from Europe. She is a fine girl as you observe. I should really enjoy seeing a good looking girl again. It was a shame not to make Joe 1st sergeant.

(for John's benefit I should have said halt) while the pioneers cleared the road of trees and rocks put there by the rebels. The entire distance was about eight miles. The march was endurable as long as we were ascending as there was shade, but coming down we were exposed to the full rays of the sun which were decidedly intense. I never suffered so much on any march I have yet made, and my experience now is not small in this line. One of my men was sun struck on the march and an other as we halted in this camp. I was good for nothing the rest of the day and all yesterday. In the evening I walked over to Gen Burnsides Head Quarters to see C.W. Loring & Mr P. Jackson who are now here with C.G.L. Jr. They saw Jule in Baltimore and say he looked very well indeed, and is very happy. It did me good to see someone from home Lieut. Hill has not yet arrived but is expected the end of this week. I had a letter from him in New York. By the way John Cobb told me a box had been sent to him by Harr[??__??]den's express directed to the regiment near Sharpsburg Md. & that it contained a package for me. This is hard luck. We shall probably not see the box for months. Never send anything by express until we go into Winter Quarters, and not then unless I write that the express goes to the place. I have not yet received my money from Adams Express, but I hope to get it through Mr Loring. I have had a servant for some time but I have so much to do when I am in camp that I keep him pretty busy and do not always like to spare him long enough to work my clothes. Accordingly sometimes when I am by the brook early in the morning taking a bath, I treat my clothes to the same, though usually, since I have had a servant that pleasure has devolved upon him. I am too becoming quite an experienced cook theoretically. My boy is not very well ported and Lieut. Pope and I have to instruct him. Many are the discussions we have as to the best mode of cooking this, that, and the other. Under our directions palatable messes of tapioca, and corn starch mixed merely with water, no milk or eggs, arise. I have also discovered that flour and water mixed together with a little salt make a batter which when fried on a tin plate over a fire made of rails is not bad. I should inform you that the substance of the fire and the pan are not altogether essential. You ask if I have not a change of clothing and I answer, most assuredly not. When we left Camp Whipple Virginia we were told our baggage would be sent after us in a few days, but we have not seen it yet and will not before we go into winter quarters. The Adjutant was sent to Washington after it the other day but could not get it. We are now located in Pleasant Valley between the South Mountain range and Elk mountain. The nearest town is Wyerton. It is two or three miles off. In Harper's Weekly of Sept 27 is a map of the region which will show you where we are. Our camp is on a side hill and I have much difficulty sleeping in comfort.

Oct 10. 11 AM

I had my tent put up last night and am now quite comfortable. About 7 o'clock last evening Major William and I went over to see Mr Loring again. We spent a very pleasant evening & took tea with them. Cart before the horse. This morning I had the ground inside my tent picked over with a pick and smoothed. Tonight my bed will be soft. How should you like to have to depend on a pick to have a soft bed. Live & learn. You ask me not to expose myself and to be careful in the next battle. But there would be very little use in this. There is danger everywhere and no more in one place than the other. One man after firing his last round threw himself on the ground thinking he would be safer there. In a few minutes he was struck in the head and instantly killed. If had had been up, he would at worst have lost a leg. I do not believe in a man needlessly exposing himself but if duty requires exposure he should not hesitate a moment. Besides to govern a company well a captain must be respected. If he exhibits signs of fear or crouches on the ground through the action he is despised. If you could hear the way the men talk if some officers who kept quiet during the fight you would not advise me to follow their example. I do not think my men would refuse to follow me anywhere now because they know I would not ask them to go where I was afraid to go myself. Enough of this. Love to all. kisses in plenty for the babies. Write soon.

Your affectionate
Brother John

In Camp Pleasant Valley
MD.
Oct. 19. 1862

My Dear Mother

Our semi-weekly mail arrived last night and brought me your letter of the 12th postmarked the 15th. Is the post office to blame for its not being mailed sooner or did you omit to send it. Four letters are too precious to remain in the Dedham office three days. It is bad enough to delay at this end of the route. I have now six letters on hand unanswered, all received since yours, some on business others from friends, but I shall give yours the precedence, though there is nothing new to write about. We are still in this valley in more senses than one. We drill four hours a day which gives me exercise enough. In addition there are a hundred and one things which a captain has to see to. Men's quarters to be inspected, clothing to be drawn and given out, letters from sick and wounded to be answered, furloughs to be made out, & many other things to be done; all which occupy my time so much that it is night before I know it. When we were in camp in Virginia Lieut. Pope suggested that we should procure a mess chest, accordingly he wrote his father for it and soon after we camp to this camp it arrived. It is extremely convenient I can assure you. It has legs which fold up inside, when it is used as a chest. In camp by extending the legs it makes a very good table three

tea pot
feet square. I am now writing upon it. With the chest came plates, knives & forks, cups, spoons,
^

frying pan, boiling pan ***[Note: The p in pan is written over an illegible letter.]*** dipper, alcohol lamp, and other conveniences for cooking. Food is plenty though the cost is very high. Bread 25 cts a loaf, pies 20 to 25 cts. Potatoes $2.00 a bushel, sweet do 4 cts a pound , butter 40 to 50 cts pound. We can buy beef three or four times a week at 11 cts a pound from the brigade commissary. He sells also splendid brown sugar at the same price. I enclose a sample. Any thing which is furnished to the men as rations we can buy of him at the price the government pays for it, but what we buy of outsiders costs a heavy price. I manage to live pretty well. This morning I had baked beans, bread & butter, and consolidated coffee. For dinner, beef steak, roast potatoes, squash and tomato pie. I have not tasted ***[Note: Indecipherable mark with one angled vertical line and three angled horizontal lines.]*** squash or tomatoes before since I left home. For four days last week I had roast sweet

them
potatoes every morning for breakfast. The desire of my great toes for liberty induced [—it—]
^

to break the bonds of restraint imposed upon [—it—] them & they suffered accordingly by contact with the leather of my shoes. [?Nor?] being disposed to gratify them in their desire I procured a needle and yarn and executed a most skillful piece of darning. When I come home I intend to take the premium at the Cattle Show. The new Dedham Company seems to be much better cared for than the old one. I hope that the wounded men of my company of have gone home on furloughs will be well cared for. The box for Johnny Cobb has not yet turned up. Cobb has gone off to Washington in detached service to pick up stragglers from this regiment. I hope he will be able to find the box in his wanderings. It is probably either at Frederick City or in Washington. More likely the former as that's the nearest point to Sharpsburg to which the express ran at that time. I have sent to Mr Wakefield for a valise and coat & many other things. Have them sent as soon as possible, for while here things directed to Harpers Ferry and sent by Adams Express will reach us. A few days ago it looked as if we should move soon but now I should not be astonished if we remained here for some time. Some troops must be kept this side of the Potomac. Still we may go at any time. My trunk has not yet arrived but I expect it tonight or tomorrow. How glad I shall be to see it again. When my valise arrives I think I shall send my trunk home. The nights are extremely cold. Two double blankets and an overcoat are hardly enough to keep me warm. Lieut Hill has returned without bringing me anything from

temporary

home. He is now in ^ command of Co k. the Roxbury Company all of whose officers were wounded in the last battle. Lieut Pope shares my tent. I enclose his photograph which please put in my book case draw, small one m=next to window. Before we received our tents we had some very cold nights. One morning there was frost on my blanket when I awoke. If there is any room in my valise put in a small bottle of liniment, Dr Stimpson's. Give my love to all. kiss the babies for me. Tell Ned I shall write to him very soon.

Your affectionate son

John.

There is a postage stamp inclosed which only needs mucilage to be good.

In Camp, Pleasant Valley, Oct. 23/62

My Fanny

I received your letter of the 15th yesterday. The postmark was the 17th so I infer there was a delay in somebody's coat pocket. I suppose you have received or rather Mr Wakefield has the letter amendatory. In primis business. Ask Mr Wakefield to pay my tax bills out of my money. Next my lieutenants shoulder straps are good ones and I think I can and I think I can change them into captain's by sewing in bars. I require four new bars of bullion one inch long by three eights of an inch wide. I have rubbed a paper over the bar [?n—?] on which will show the

a

pattern. The centre is sunk a little with a [—two—] ridge[—s—] on each side and a beading outside. I bought my straps at Pollands on Court St, [—I—] I think Mr Wakefield will have no difficulty in having new bars made there. If he can please send them by mail. This will not

for Officers

affect my order as to the straps. In the next place I should like the U.S, Digest of Tactics ^ by Capt Staffen published by A.K. Loring. Send this also by mail. The 9 months men seem to be more careful than the three years if the order is carried out that you sent me. We received orders before the last battle to have our swords drawn and to cut down the first man that should run. This is about all that swords are good for. Officers don't often cross swords with the enemy. How an officer can avoid being known I do not see unless he takes his whole uniform off and dresses as a private. If he does not command the full and entire respect of his men he has no influence in his company and can accomplish nothing but by compulsion. Now if in battle he does not do his duty and his whole duty, if he does not show the men that he is not afraid of being hit, he cannot expect them to do their duty & expose themselves when necessary. An officer must say "Come with me, and not "Go there." He must lead and not follow. I received tonight Monday's Advertiser containing an article on the battle of Antietam. It was written I think by Mr C. W. Loring who has lately been out here. It contains several mistakes however. The Rhode Island regiment is not in this brigade. The rebels had no rifle pits to defend the bridge. The article is in the main correct. We have been drilling constantly since we came to this camp& when we go into the next [—camp—] fight we will be able to do as well as the rest. I thought yesterday that we might winter here but now it looks as if we might move very soon. An order was read tonight to the effect that no more clothing was to be drawn for the men in the <u>impending</u> <u>movement.</u> The order was general for all McClellan's forces and was not meant specifically for us therefore if only shows that there is to be a movement soon and not necessarily that we are to take part in it, though we probably shall as Burnside has been ut over the 2nd and 12th corps as well as the 9th which we are in, and part of his troops are already across the Potomac. I am in no hurry to leave our present quarters for it is very pleasant here, and I am quite comfortable. I can get fresh bread every day although it costs from 20 to 30 cts per loaf, and now I have a jolly bedstead with plenty of straw, and a tent heater by a patent furnace. The fire is outside the tent. A flue runs through ^ under ground and turning at a right angle near the door runs under the side of the tent for some six feet to a chimney constructed of a beef barrel. Over the flue are flat stones which are soon heated & the tent is quite comfortable. Lt. Pope was sick one day & being half scared to death wanted me to write a letter for him which I did. The next day his brother came to see him & he recovered. If you want to understand how a person can sleep on the ground & eat flour & water mixed & fried, walk out from Boston to Dedham on one day when the thermometer is about 80° without anything to eat. You would be glad to sleep anywhere and eat anything. A soldier's or rather an officer's life is not so bad except when he is on a forced march & then it is really rough. Love to all. kisses for the babies

Your affectionate brother
John.

I will write to Jennie soon. The money came safely in Tom's letter. Pay rolls are made out but we may not get any pay for some time. In Camp near Wheatland Va

Nov. 1. 1862

My Mother

I wrote to Ned on the 29th ult. and gave him an account of our march from Pleasant Valley to camp near Lovettsville. That afternoon we left camp and march about eight miles, then turned in after dark. I had just gotten my fire started and was preparing to pitch my shelter tent when orders came for the three companies on the left to go a little further on and encamp near a battery in order that we might be able to support it if necessary. The new location proved to be much more desirable that the old. It was in a wood, the battery being put outside. I turned in early but was aroused at one o'clock by orders to have my men ready to march a daybreak and to draw rations of meat at once. This put an end to sleep. We did not succeed in getting the meat at once, but by three in the morning every one was busy cooking beef steak in a variety of way. Some on frying pans, others in tin plates, others on sticks, and others on the coals without the [?interoperation?] of anything. We were all ready to leave at day break but did not start until half past seven. After marching for two hours we halted at our present camp. It is near Wheatland [—W—] Va. I do not suppose you can find the place on any map but it is between Lovettsville and Hillsborough. The country around here is very beautiful. Our camp is on a bank some fifty feet above a creek. The banks are lined with trees and the large oak trees around make the camp by far the most pleasant we have yet been in. The place does not seem however favorable to sleep. The first night I was up until two o'clock making out pay rolls. Last night, I was congratulating myself that I should have a chance for a little rest but at half past five our company was ordered to picket duty. We went about a half mile from here. Other pickets were beyond us about two miles out so we apprehended no danger. At half past ten Gen. Ferrero rode down and told me he had intelligence which led him to believe an attack would be made in the night. I got my men all ready and advantageously posted. About half past twelve cavalry was heard approaching from the direction of the enemy. When they came up they were halted in proper form and the countersign demanded. This was given & they were allowed to pass. I knew by the noise they made coming up that they were our men. The officer in command told me that two mile further out they frightened a picket of two companies nearly to death. He said the men ran in every direction. These companies belonged to the 163rd New York. The night was a beautiful one and everything passed off quietly. In the morning I returned to camp and found nine letters awaiting me. I can assure you the night was a most pleasant one for I have not had a letter for a week and this was the first mail of any consequence since leaving Pleasant Valley. I had a letter from you, one from Jennie and two from Tom. The others were mostly from soldiers in hospitals. There has been some heavy firing this morning and I think we shall move soon. Perhaps tonight. If we do not, then, tomorrow. The exact purport of our expedition I have not ascertained. From our present position we can either go through the Blue Ridge of the kittoctan Range. We have been approaching the Blue Ridge, and I learned today that heavy forces were moving from Harper's Ferry, and I think we shall join them & attack Winchester. The campaign must necessarily be a brief one; and if my valise has not yet gone hold onto it until I write. Let the bars of my lieutenant's shoulder strap be sent by mail. I forgot to say that I also received the book [—on—] called Digest of the U.S. Tactics which I wrote for. It seems to be a very good book. I will keep you informed of my movements as well as I can but if you dont hear do not be alarmed for the mails are very irregular. I am in splendid health and although I have had but little rest for three nights past I scarcely feel it. How glad I am that Edward's body has reached home. Did Mr Wakefield receive his watch which I sent by Magen Fay of Chelsea and my rebel edition of Tactics which I sent by Mr Bartholomew of Roxbury. The mail is about to close. Your affectionate son

John

Fauquier White Sulphur Springs
Nov 14. 1862

My Mother

Just got in from picket and tired out but I must write a line or two. I had sent a long letter to Jennie, eaten my dinner, mended by stockings – darned them – and was just commencing to read the papers received by the morning mail when a cannon was heard across the river in the direction of Jeffersontown – Our regiment was at once order out when to my consternation I heard that Lt Col Carruth and Adjutant Wale's [—were—] had gone across the river to get dinner & were probably prisoners – All sorts of rumors were about but as I afterwards learned the fact I will not now stop to relate them but will merely state the facts in their order – We marched down to the bank of the river & lay for some time under the shelter of a bank to defend the ford of the river. About dark we were ordered across the river. Before we went we could see the rebel cavalry on the hills less than a mile from the river. We went across, some companies deployed as skirmishers and advanced beyond the hills above mentioned. Strong pickets of cavalry were thrown out and infantry from our regiment. My company and two others the whole of my command were held in reserve in the rear of the pickets. There was no disturbance during the night and this morning we were relieved & got back to camp at 1 P.M_ Yesterday our pickets of cavalry extended to Jeffersontown about three miles from here. I regiment of rebel cavalry & a piece of artillery came out and attacked them. There was considerable skirmishing but no loss on our side though the pickets were all [—deter—] driven in. Our cavalry took possession of Jeffersontown this morning again_ Now for the capture_ To the right of the road running from here to Jeffersontown & across the river is a house occupied by Mr Miller a secesh_ The adjutant had been there the day before yesterday persuaded the colonel to go with him for dinner. The hime is not more than a half mile form our camp & as our pickets were out more than two miles beyond no danger was apprehended. The went, sat down to dinner, but before they were through the house was surrounded by fifiteen or twenty rebel cavalry under command of Gen Stuart, and they were prisoners. At the same time they were captured two men in the 51st Penn – and one of our cavalry who was sent to warn Col Carruth of his danger. At first I thought a trap had been lain but this morning I visited the house & was convinced that the man though secesh was honest & had no hand in the matter. I went to an other house in the neighborhood & got a good breakfast of hoe cake & fresh milk. I ran some risk in doing it as this house was beyond our pickets, but concluded that the rebs would be satidfied with the prize already obtained. A large & fine home near is riddled with cannon balls. A small tobacco drying house had five balls in one side. This occurred during Banks retreat last August. The loss of our Lt Col is a most unfortunate thing for us as he was a splendid officer – cool & brave & beloved by officers & men.

Your affectionate son
John

Camp near Warrenton Junction
on Alexandria & Orange R.
Va
Nov 16. 1862

My Mother

Only time to write a few words_ Yesterday as the Col had been taken prisoner, about which I wrote you, Major Willard took command & appointed me acting major. My letters are however to be directed as before_ We left White Sulphur Springs yesterday morning, the baggage train of the whole division about a mile in length following – our regiment was in the rear of the whole division following us was the long ambulance train & then the wagons We took the road which [—leads—] goes by the hotels & then turns to the left along the river — As we approached the river I saw our cavalry pickets rushing down the hill on the opposite side & suspected danger. In a few minutes & at a time when the whole ambulance train was in full view the rebs planted a battery behind Hoffman's house on the other side, the house I wrote about a few days ago, and the shelling commenced. Our regiment being near one was order back at once to protect the train. Back we went & for fifteen minutes were exposed to a hot & heavy fire from three pieces_ Only one of our men & he not of my company, was injured. Most of the baggage train was turned back & went by an other road_ For a time the shells flew lively over us, around us & on each side of us. I made up my mind some time ago that I would not duck my head in the event fright I should be in & I don't think my head went down once. After the train passed [—the—] we were ordered to lie down behind a hill to support one of our batteries on the hill. There though our of sight we were not our of danger for a great number of shells struck around us, one within a few feet of the end of the regiment_ Last night we slept in the wood not far from the seat of action & today we march to this place which is about ten miles from where we were yesterday. I should have said that when the rebs first commenced firing they shelled a company of the 7th R.I. which was on picket at the ford, out & then charged with cavalry & attempted to cross_ The firing was too heavy for them & they had to retreat_ Our batteries after a while silenced the reb battery & the skirmish was over. Two of the wagons in the train got stuck in the mud in attempting to cross a brook & were burned_ A lieutenant of one of the batteries named McIllvaine & a private were wounded seriously_ I heard of no other loss_ I am very well & being major have not nearky so much to do as before_ Lieut Pope is in command of my company_ Write soon

Your affectionate son
John

Camp near Falmouth
Nov 21. 1862

My Mother

I wrote to you a few days ago that I was acting major of the regiment. The honor was but a short lived one. Gen. Ferrero send down an order that the two senior captains should be appointed as acting Lt. Col. and major& I am now therefore back with my company. We reached this place which is directly opposite Fredericksburg, Va. day before yesterday in the rain. Since then we have had nothing but rain. Everything is damp and uncomfortable. Nothing is more dreary than rain in camp. & I shall be delighted when the sun comes out. It is said that Burnside's intention to spend Christmas in Richmond. I trust he will do so & that I will be with him but am afraid the mud will be in the way of progress at present. I am very well. I

e

shall keep you posted in regard to my movements as often as possible. Th[—is—] frequency of my notes must be my excuse for their brevity. Love to all

Your affectionate son
John

Camp near Falmouth, Va.
Nov. 27. 1862

My Dear Aunt Jane

Your letter of the 17th reached me on the 21st and since then I have been trying hard to find time to answer it. I am very sorry to hear that you have been sick and hope with all my heart that you will enjoy good health for the rest of the winter. Poor Joe had a hard time of it by all accounts before he got finally off. How I longed to see him & bid him God speed. I am glad that Mother has someone with her this winter. It will be so much more pleasant for her than
at
being alone. I did not ^ all relish the idea of her being alone. I have plodded many a weary mile. Just one month ago today we left Pleasant Valley, Md. And since then have been almost continually on the march. Up hill and down through snow and mud have be tramped but I have stood it well & never was in better health. While nearly every man in the regiment is barking away as if he was a candidate for the grave yard I have nothing to complain of but a very slight cold in my head. Now don't you say again that I never can stand it. I have stood it as well as any man in the regiment be he private or officer. As you have seen by the papers we are now opposite the city of Fredericksburg and before you receive this an attack will probably be made. We have here now about one hundred thousand men & the rebs are said to be in strong force across the river. This is Thanksgiving day at home and has been observed here so far that we have had nothing to do. Today I had beef steak and hard tack_ Don't you envy me I hope that by an other Thanksgiving I may be at home and that the family may be united, and the country at peace. Old regiments out here begin to think that they will have to stay out theor terms of three years but I do not permit myself to despond yet. That there will be a great battle soon is I think certain. If anything does happen I will write to you as soon as possible but do not be alarmed if you do not hear for often there is no chance after a fight to send a line for several days. Remember me to all my friends_

Your affectionate
Nephew
John

(Letter fragment)

letter of the 16th giving an account of the fight. That of the 17th was sent by Mr Carruth Uncle of the Col. who joined us at Sulphur Springs on the Friday before we left. With him was Miss Gilson an adopted daughter of Thayer Fay of Chelsea. She is a second Florence Nightingale. Her whole time is devoted to the army. She does not content herself with assisting the hospital in some large city but goes where she is most needed to the hospitals near battle fields. She was with our brigade hospital soon after the battle of Antietam and Pleasant Valley. A hospital is a sickening place, not so much on account of the wounds and blood and filth, but because of men uncared for and neglected. I could not help repeating "A ministering angel thou," when I saw Mss G. in a hospital near Antietam, at night making tea for the wounded and soaking their hard bread in milk. The scene was one for the painters. The hospital was a large barn. The floor covered with straw and on it the wounded and sick groaning with pain. In the center Miss G. dispensing her bounty with liberal hand. She is a trump and if it is my lot to be wounded may I fall into her hands. I came to the conclusion that woman is an institution. To return to the morning of the fight. After our regiment was a battery, then a train of ambulances and then the wagon in which Miss Gilson rode and after that the wagon train of the division. The greater part of the latter were turned back but all before it had to pass through the thickest of the shelling. Miss G. behaved splendidly. She shamed some of the men. The 35th received more praise than it deserved. It did not repulse a greater force of enemy at the ford, because the rebs did not come to the ford while it was near there; nor did Gen Sturgis have a ^ shot under him. The wagons that were destroyed were lost partly by the stupidity of the drivers getting them into ditches and partly by the stupidity of some one who set fire to them when there was no need of so doing. If you have not received an account of the skirmish I will give you one hereafter but I have written so much now that you must excuse it at present, nor can I give you an exact account of my daily life since because nothing of interest has taken place. Here are some dates if they will be of service. (END)

Camp near Falmouth. Dec. 5. 1862

My Mother

Yesterday afternoon while on battalion drill an order came to Gen, Ferrero to send a regiment to support a battery. According to our usual luck we were selected. Pack up was the word and at it we went. By half past four we were on the march. After going about two miles we halted in a field and the major having lost the way sent back for a guide. After waiting an hour the guide arrived and on we tramped for a mile further. We are now on the extreme left below Fredericksburg, supporting a New York battery of four Parrott guns. This battery has been here entirely unsupported since the 20th of last month This is rather singular If the rebs had know it I believe they would have taken the battery without much trouble. The paymaster was here the other day and paid off all the regiments in our brigade except ours. On the day assigned to us he was ordered to Washington & so the 35th has to wait. Our luck. _ There is a rumor that the 9th Army corps is going to remain here to guard the railroad between here and Aquia creek. This is founded on the fact that while the other corps are receiving shoes & clothes we have nothing as yet. I believe however that we shall advance with the rest. Love to all – I wish you would send me soon by mail a small pocket diary for 186[—2—]3. The mittens for my men might be sent by mail by doing them up in two or three bundles & leaving enough open to show what the contents are. The postage would not be much.

Your affectionate Son
John

P.S. I want you to take $2.00 of my money & give Christmas presents to each of the four babies for me.

JL

(Letter fragment)

Camp near Fredericksburg, Va
Dec. 19. 1862

My Fanny

I must try to write you a few lines though I do not feel much like doing anything, You have heard from the papers all about our defeat, for it is nothing less; how we crossed the river, fought on Saturday and retreated before daylight on Tuesday morning. I received your letter the other day and this morning came Jennie's. Our brigade was in the fight [—Friday—] Saturday. All Saturday night I sat up with Major Willard in the hospital and got no sleep. Sunday night I turned in at 6 but had not got to sleep when our brigade was ordered to fall in. We had to go the extreme front and do picket duty. But little chance for help here. All day
Mon
[—Sunday—] we had to lie there no being able to move twenty feet from our position without getting a shot from the rebs. [—Sun—] Monday night we were relieved & crossed the river after midnight, marched to camp behind the batter I before wrote about & arrived there at three o'clock more dead than alive. Tuesday we rested. Wednesday morning marched three miles and joined the brigade at its camp. The day was bitterly cold. I had just got my tent pitched and was trying to start a fire when I was informed that I was detailed to command a burying party from the brigade to bury the dead on the other side of the river. Under ordinary circumstances I should have declined as I was worn out but I wanted to find Lieut Hill's body if I could. Accordingly I went and did not get back until eight at night. I was almost frozen. The scene was a horrible one. Most of the bodies were stripped to the skin. My squad of 50 buried 115. Six hundred were buried during the day. Hill's body I could not find. All of the bodies on the part of the field where he fell had been buried before I arrived. He had sent home $250 or $200 a few days before by Major Fay so he did not have much money with him. What he had undoubtedly fell into the hands of the rebs together with his sword and revolver. His valise and other things I will send home to his mother as soon as I can. Yesterday I found myself all used up and today for the first time since I came out had myself reported as sick. I knew if I did not I should have some duty to do which I had rather leave undone. I [—send—] write to Mother last Sunday stating that I was safe & after directing it, put Tom's direction on in pencil, enclosed $200 & sent it by Roberts who went with Major Willard's body to Washington. He has returned & says he gave it to Mr Fiske the son of A.H. Fiske. I hope it will arrive safely. Tell Tom if Mr. Parsons has not paid up to [—depos—] give $50.00 of it to Mother as a Christmas present. If Mr P. (END)

Happy Christmas to you – I suppose it cant be merry one

Dec. 23rd /62

[/Your shoulder straps are done – shall they be sent by mail?/]

John Lathrop shoulder insignia front.

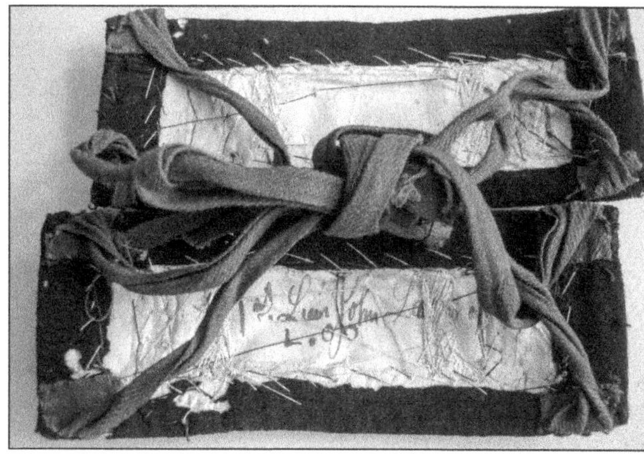

John Lathrop shoulder insignia back.

Dear John

We have been anxiously looking for another letter from you, but as none has come, we take it for granted there has been no mail from the army. I am thankful you sent your letter as you did, for it would have been terrible to wait all this time without hearing from you. Do say in your next letter whether you were hurt at all, for though I suppose you would have mentioned it, if you had been, still I should feel easier if you would say that you were not. You cant imagine how terrible the suspence was till the list came. We heard [—Wednesday—] Monday morning of Willard's death, but no other names were given in the 35th till Tuesday. Your letter was the first correct list of your company that came.

People are dreadfully blue about the war, it seems awful to lose so many men, and yet accomplish nothing, besides everyone supposed the Banks was to attack Richmond from the rear, and there is a general feeling of disappointment that he is not going to do so.

We have heard nothing from Julius since he left Fortress Monroe, whether he went to Newbern or not we cannot tell; though Joe's last letter said the 38th was to be in their division. I rather think they both were in the attack at Kingston, and now have gone to Goldsboro. Nothing has been heard from the 23rd since the 10th. I do hope we shall get letters soon. Mother bears all this anxiety much better than I thought she would. I believe she was right in keeping home this winter, after all; the little everyday duties that she must attend to, prevent her from giving up her whole time to thinking of the dangers her boys are in. I believe if she had had nothing to distract her mind last week she would have had a fit of sickness. She and George will spend Christmas with me, and perhaps Aunt Jane – I really dread Christmas, it will seem so lonely; we shall have no tree this year. Do you remember last year how you planned to have a tree at Mother's this Christmas? I am very much obliged for the childrens' present, it was very kind in you to remember them, Johnny will be astonished to find a present from "Uncle John", he says he means to keep the book you gave him last year, to show you when you come home. Frankie is very well, he weighs 17 ½ lbs. so you may imagine he has grown since you saw him.

I had a letter from Mrs. [?Massiah?] the other day, she says she has a photograph of Julius and of Joe and she wants yours. Have you one to send her? I wish you would have some now taken as soon as you have a chance, in your uniform.

I had almost forgotten the message I was to give you. Bunker's father (or brother, I don't which) wants to know if he had been paid, and if you know any thing about the money, and if there is any chance of recovering the body.

What was the reason Willie Titcomb did not get that situation as clerk. Mrs. Montague said you were very kind in trying to get it for him, and did everything in your power; they were quite disappointed he did not succeed_ He seemed to think the reason was because he belonged to the 35th – was it really so? Do write soon – you see I wrote two letters to your one, so I hope the answer will be a long one. Johnny says I must write a letter for him. Fanny.

I wrote to Jennie for some package stamps which I should like to have as soon as possible

Camp near Falmouth, Va
Dec. 27. 1862

My Fanny

Your letter arrived this morning. I wrote to you about a week ago and presume you have received it before this. Christmas was very pleasant here as far as the weather was concerned. We had nothing to do until dress parade. Col. Loring called the day before and gave me a can of turkey put up in Boston; with it I had onions and potatoes. John Cobb had a box from home a few days before with a mince pie among other things. He gave me a piece & some apples, so I fared quite sumptuously for these regions. As to myself if it is any satisfaction I can assure you that I was not hurt in any way in the late fight; and what is provoking I can not show even any damage to my clothing. In the Battle of Antietam David Sullivan who was hit in the last fight in the hand, had a bullet lodge in his canteen. What a nice place to have one. No such luck for me. I sent my old coat home by Mr Fisk. It looks as if it had seen service. It has been with me or rather on me in every fight I have been in. Four times under fire in as many months. I thought I wrote to you when I sent my shoulder straps on to have them sent by mail as soon as done I have been expecting them every day. Again I wrote to have a pocket diary sent out by mail, but it has not appeared. There is a camp rumor tonight that the brigade is going to New York for garrison duty – [?fudge?] – If we had kept Fredericksburg & our army advanced, our brigade would have remained as provost guard. This would have been jolly. Che sara, sara. As to Billy Titcomb_ He was rejected because the brigade quartermaster thought he did not look strong enough. It was not because he belonged to the 35th because the quartermaster wanted a man from our regiment. I trust you passed a pleasant Christmas, and I hope by the next we may all be together. I have not heard from Jule or Joe for a long time, nor do I see as yet any account of the number killed or wounded or any list. I wrote a long letter to Jennie a few days ago giving a detailed account of my doings for some time back. Since then nothing new has transpired. Mr Pope, Lieut Pope's father is here tonight and will take this to Washington tomorrow. I send by him also – Bunker's things, which he will send by mail. There is a later Harper's Weekly with a topographical map of the railroad from Fredericksburg to Richmond. I would like it if it can be procured. I have gained 30 pounds since I came out. I now weigh 153 pounds. Love to all

Your affectionate brother
John

(Letter fragment)

in his hands. The houses were wrecks. All in this part of the city were shot through and through. Soldiers were cooking in the kitchens, flour barrels lying around loose. Trunks were broken open. There was a soldier with a bonnet on his head, there another cutting out a figure from a fine engraving. In an other room there was an officer playing on a Chickering's grand, while just below the body of the instrument was a hole where a round shot that had passed through. In the entry of one house was a pool of blood showing that some poor devil had met his fate there. Trophies were abundant but I contented myself with a pictorial Blue Beard for Neddie and the last days of Pompeii for present reading. No movement was made that day. At night one of my men brought me a straw bed from a house and although out doors, I passed a most comfortable night. The next morning Saturday I was again acting major, more than half of the officers were sick in camp. We started off about nine and marched up this principal street of the city nearly to the rail road depot. The rebel batteries were behind the depot about a third of a mile & we were in direct range. Sheltered by the houses we remained here some hours. In one house I discovered a copy of The Berber and seating myself in the parlor read several chapters when a round struck the roof of the house directly opposite. There was no safer place however & I remained. Before we left the house opposite had been struck several times and the one I was in had received at least six compliments of the same kind. It was a small house. The front door opened into an entry about five feet square. From this opened on one side the parlor and on the other the dining room. Back of the parlor was a shed & back of the dining room the kitchen. The rear of the house was towards the rebs. One shot struck an
in the partition
oaken beam^ between the kitchen and the dining room but did not come through. Thinking that the next might take the parlor I took my position in the entry midway between, but I was mistaken for soon a canister shot entered the dining room and scattered small iron balls about the room in abundance. Not caring to play marbles I let them alone. Soon another shot went through the roof and struck a beam over the ceiling of the parlor. The plastering garnished the carpet but this was all. Another struck the roof and scattered the shingles over the men in the street. At length the order came to fall in. We marched beyond the railroad turned up the next street and were soon on the field. Crossing the railroad we formed in line of battle and with a cheer advanced. The position was this, though I knew but little of it at the time. About a third of a mile from the depot was a range of hills on which were strung earthworks and rifle pits. This was in front of our line. On the left was an other range and ^ the right the first range curved so that the rebs could not only fire at us from the front but also from either side. Carleton the Journal correspondent gives a very good account of the battle only he makes the position worse for us than it really was. He says it was this, A B. & C being the batteries and D our line. It was I think more line this The rebels you will perceive had a cross fire on us.

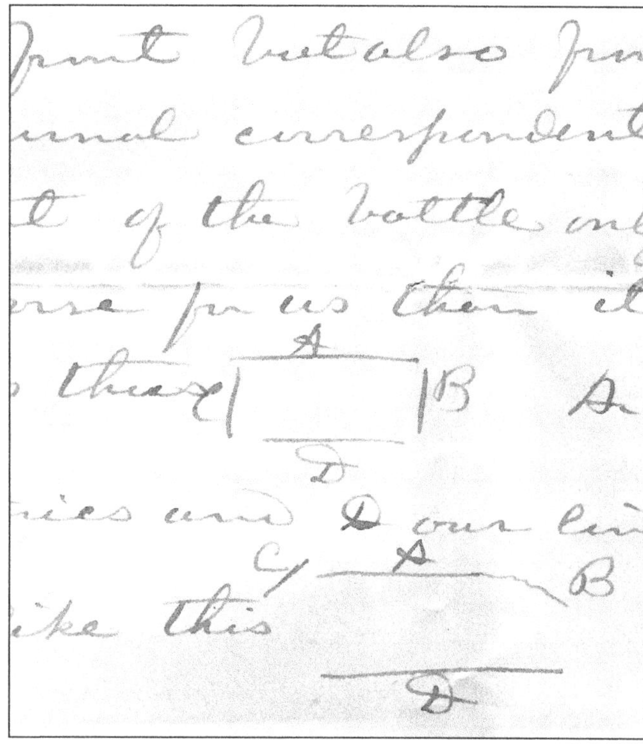

The page with the two maps that John references.

The railroad ran to the left of the works and a highway ran through a deep gully [—nearly—] which continued nearly to the furthest point we attained. The field was not more than a quarter of a mile wide and upon it was turned the full fury of the rebel fire. It was terrific. Such a shower of shot, shell, and bullets I never experienced. Antietam was boys'play to it. There was nothing to do but keep on. Fortunately about two hundred yards from the batteries was a slight depression in the ground. Here was [—comparative—] partial safety, and it was this alone which saved our army from utter destruction. To attempt to carry such works at the point of the bayonet was in my opinion madness. Here too was demonstrated the uselessness of artillery for attack when opposed by riflemen, and other artillery. The rebel artillery was protected as I have said by earthworks. The most of our was across the river in a safe enough position but utterly useless as it could not fire without endangering our own men. The batteries on the Fredericksburg side could not take positions in the open field without being annihilated. Consequently [—the—] we fought almost unsupported. Major Willard was struck by a minie ball before we reached the hollow spoken of. Lieut. Hill was hit by a shell in the leg soon after we went on the field, first taking him to the gully & afterwards to the hospital. I remained with him all night & left him in the morning to look up the regiment. Before I returned he was dead. He had on an armor vest which did not (END)

[Note: This may be the letter, referred to in a subsequent letter, that John Lathrop wrote on December 24 to his sister, Jennie.]

Camp near Falmouth, Va
Jan. 18. 186[—2—]3

My Fanny

I received your letter of the 13th last night. Day before yesterday we had orders to be ready to move at an early hour the next morning. Accordingly all was packed up and here we have remained in a state of readiness ever since. No order as yet has come for us. Adj't Baker was over here yesterday & said the 18th had received the same order & then the time had been change to 10 'clock today. A man from the 20th was here this morning & said they were ordered to be ready tomorrow morning. The delay is owing doubtless to the extreme cold weather. The last two nights have been intensely cold, days moderately so but pleasant. We shall doubtless have warm weather soon and then the movement of some sort will probably be made. Where I don't know, but have an idea we shall cross the river above or below here. No winter quarters yet. Lieut. Atkinson of Co. A of Lynn is going home tomorrow discharged on account of disability I shall send this by him and also a parcel if I can get him to take it. All my old letters will go in it. I want these kept in my trunk, till I return. From your letter I should not judge that you had received my last when you wrote. I wrote to you on the 27th of Dec. to
& on the 10th of Jan
Jennie on the 24th, ^ to Ned Jan. 2, to Mr Babcock Jan. 3. To Mother on the 5th & 6th. When my valise is sent mark the bottles cider & they will come all right. I dont believe anyone would open the valise though privates' boxes are sometimes opened by the provost marshal. If we go into winter quarters I shall most assuredly try & get a furlough. My letter to Mother of the 5th was sent by the sutler to Washington. This accounts for the short length of timeit took for my letter to get home. I have not yet received the Witness. Kerr's account of the fight is very good. Did it ever occur to you that Orpheus C. Kerr means Office Seeker? I dont want my valise yet but the mittens had better be sent as soon as you see by the paper that this movement is over. I want a tooth brush & a pair of socks sent by mail at letter postage as soon as possible. Dont wait for the result of this movement. I spent an hour darning a big hole the other day. Mr Kemp has not yet turned up. Joe I have not heard from for a long time. On the 16th I had a long letter from Jule dated the 3rd. He was then in camp about four miles by land from New Orleans. He expected to stay there some time. Tell him I received his letter & will write as soon as this movement is over. I have just heard that Franklin's Division is fighting today on the left & that we shall probably go in tomorrow on the right. I give this as a rumor. In the parcel I shall put a pipe which give to Ned as a contribution to his tower. It is a relic of Fredericksburg. I must now close but if I hear anything more [—will—] before this goes, I will write it.

Your affectionate brother
John

Sullivan's box I had to open & give to his friends when the order came to move. The letter that Mr McKew wrote to Weston H. Keene, Sergt. Major of the 20th was opened by the postmaster of the regiment & sent to me. He says there is no such person in the regt. Keene must be Sergt Major of some other regt.

I have been out & got a little additional light. Our sutler has just come from Washington. He says that Franklin has crossed with his whole force on our left & that Hooker is to cross to the right while Sumner's division holds this position & is to act as a reserve. Our surgeon, Dr Lincoln, says that Gen. Nagle who commands our division in place of Sturgis who is temporarily absent, told him that Sumner's division would not move at present, but would be ready to move to support either division. The idea is that as Longstreet has taken away a good many troops to the West this is our chance to strike a blow.

Answer to No 1.

Camp near Falmouth, Va.
Jan. 23rd. 1863

Dear Jennie

I wrote a long letter to Ned the other day. Commencing on the 20th & ending Wednesday morning rather abruptly. Tuesday [—morning—] night it poured in torrents and the order to fall in at 4AM was countermanded. I was quietly reclining on my bed of barrel staves when the sergeant major stuck his head inside my tent and said that the line would be formed for picket duty in fifteen minutes. I informed his that I could not see it, and I did not for the line was not formed in a half hour. Up I jumped dressed, swallowed a cup of coffee and a few hard tacks and without a chance to wash my face or brush my hair started off. The mud was jolly and deep, and of every variety and color. Before reaching the river there is what is called in these regions a run. This in ordinary times is fordable and we can cross without wetting the feet by going across on stepping stones, but on Wednesday it was far different. The run there was over four feet deep and rushing along like a mill stream. One individual tried to jump across and landed in the middle up to his neck. The 21st Mass was with us and the Lt. Col. emulous of Julius C. "plunged in and bade us to follow" So indeed we did not for down went the L.C. until only his head was visible, and out he scrambled on the other bank a sadder but wiser man. Some of the 21st went through with better success but not being desirous of distinguishing or extinguishing myself I declined to follow suit by wetting mine and went up the stream in search of a bridge. I finally found a board across in one place and crossed on it dry shod. We went to our old ravine near Falmouth but it was cheerless enough. Everything was soaked, and the rain was still coming down. After the pickets were posted Capt. Gibson, Lieut Pope & myself took a tour of observation in Falmouth. Everything was dirty and uncomfortable. No sidewalks and mud, mud, mud. Finally I finished up the morning by going into a home and sitting by the fire two hours. There was a negro woman in the house who belonged to Fredericksburg and who was there at the time of the bombardment. The accounts she gave were amusing. She said most of the inhabitants had left the city sometime before and were in camp outside. A few days before the bombardment the weather was so cold that they could not stand it and came back. How pleasant it must have been for them. – Dined on an apple pie and a cup of coffee and spent the afternoon in an other house by a good fire. To do the people justice in the South so far as I have seen there they are always willing to let any one sit by their fire and partake of such food as they have. Nor do they do this sullenly as if there was no choice, and Union men and secesh are alike. The prospect of spending the night in the ravine was not pleasant and I looked around to

see what I could do. On the bluff near the ravine is an old brick church, wall roof and windows in good condition, but no seats or fire places. Into this the left wing of the regiment was turned, and there was a fair prospect of passing at least a dry night. The good people of Boston are highly indignant of the British for using their churches in times gone by, but our army does no better. Soldiers must be kept comfortable if it is possible. They did it in this instance by knocking off plaster from the walls and making a bed [—off—] of it five or six inches deep. On this w warming fire was built of laths and what wood work was handy. Near the church was a small white shanty a story and a half high. Down stairs I found four men of Franklin's division who had not been able to keep up. They had a good fire in an open fire place and defied the weather. There was plenty of room for an other bed and the three officers of the left wing turned in. Yesterday morning we waded back to camp through the deep mud and rain not much the worse for picket duty. Today Franklin's Grand division has marched back and the grand movement is over. No bridge was laid & no troops crossed over though if it had not been for the rain there would probably have been lively times before now. Winter campaigning is an absurdity. If there was any certainty of two weeks good weather something might be done. A rain of a few hours makes the roads knee deep in mud & loaded teams can not get along. The sun is now out but nothing can be
even
done for a week of two ^ if we have good weather all the time. Some of the troops say that on reaching the place they expect to cross a board was seen on the other side with the inscription "Burnside's played out;" and that a reb officer asked them if they wanted any assistance in laying the pontoons. Certain it is that the night we were on picket the reb lights & fires were up the river about in the direction where the army was expected to cross. Now to answer your letter of the 18th which I received last night. On my return from picket yesterday morning I found a package containing the soap Ned sent. I am extremely obliged to him for it. I tested its merits. It is as the label denotes superfine. Mr Kemp with my shirt has not yet turned up. I hear however that he is somewhere in the regions and I trust one day to see him. Alas living on marching rations has taken down my rotundity and my face has grown thin again. Today I had however a dish of ham and eggs, latter purchased at a rate of 65 cts a dozen. The photographs you sent are admirable. Neddie's expression is perfect. How I long to see him and you all. I have written to Mrs Willard, Sen. From the letters I have receive d from her and the Major's wife I should think the Mother took her afflictions the harder. So Harry Fisher is still in Dedham. Where is Randolph Clark now? How well I remember the last theatricals the Clarks had and how I enjoyed them. I wonder when I shall see a theatre again. What a provoking creature you are to ask me if I would not like to know what are the ideas of a certain young lady on the subject of marriage, more especially as you say you think they are right ideas. Of course I should and in your next I shall expect them verbatim. As for my marrying that is a thing I never expect to happen. Who would have me who would be worth having. There are wives to be had

by the score no doubt but the few choice ones are beyond the reach of a poor Attorney at Law who has had work to make both end meet in the state of single misery. – I had a long letter from Jule dated the 3rd inst but Joe I have not heard from for a long time. I should like a copy of [—Joe—] the picture of Joe and July but I am afraid it is too large to carry about. If you can get it in a letter envelope please send it. If not have a copy taken and saved for me. Where is Ned Sherwin now? Sue Sohier I should like to see very much when she makes her visit to Glen Ridge. Remember me to her. I am glad to hear that Joe has the true grit, but the more I see of this war the more I am convinced that gentlemen make the best fighters. Noblesse oblige. To a person of any musical taste the hum of the bullet and the whir of the shell are not in harmony. The tenor and bass are too mingled. I am expecting a letter from Mother every day. Lieut Brooks has resigned on account of disability and goes home tomorrow I shall send

for Ned

this by him and also some natural leaf of tobacco ^ which I procured at Amorville, Va. on the 9th of Nov. last. It will do for the pipe I sent Ned a week ago. Write soon.

Your affectionate brother

John

(Letter fragment)

Camp near Falmouth, Va
Jan. 28. 1863

Dear Aunt Jane

I received your letter a day or two ago and hasten to answer it. It is hardly fair to reproach me for not writing to you and sending all my letters to Dedham, for I have answered every letter you have written to me at once, and moreover I knew you were at Dedham and would see all the letters I wrote. My letters are common property and merely directed to one or the other of the family because the must be directed to some one & because that person has just written to me. You have probably all been worrying at home on account of the abominable reports that were in the papers last week. A week ago [—today—] yesterday Franklin's and Hooker's Grand Divisions moved & got stuck in the mud. Sumner's Grand Division in which we are, was to start at four o'clock the next morning. All night the rain came down in torrents, the mud in the roads was knee deep and nothing could move. The troops had to move back to their old quarters and things are as they were. No bridge was laid & no troops crossed the river. Nor can anything be done for some time. Yesterday it rained hard & today we have had a driving snow storm, snow six inched deep. Winter campaigning is humbug. I met Major Stephenson the Sunday night after the battle of Fredericksburg when we were on the battle field relieving picket. The 32nd is not more than two or three miles from here but I have not yet been over. Burnside has resigned & hooker is now in command of the Army. Sumner and Franklin have either resigned or been removed I don't know which, and every thing now is in a mess. Hooker is a splendid fighting general with a small body of troops say from 10 to 20000, whether he will do equally well with 100000 remains to be proved. I was very sorry to hear of Mr E. Eliot's death. The blow will be a severe one to his family. O do not suppose that he has left them very well off. Do not be alarmed at any rumors you may see about movements in the army for nothing can possibly be done for some time. I am very much obliged to you for the socks and will be still more pleased when I can have an opportunity of wearing them. (END)

Jan. 29 Pleasant today after a furious storm all night. No mail today from home.

Camp near Falmouth, Va. Jan. 28
1863

Dear Mother

It is no wonder my letters come home in various times, for the conveyances are very varied. If any citizen appears in camp I pounce on him & if he is going to Boston he has a letter to carry. So if any one, our quartermaster or sutler goes to Washington he too is burdened with a paper missive. Again the only mode of communication is through the agency of my respected employer, Uncle Samuel who though always slow is not always sure especially when there are war[—s—] or rumors of wars. And, speaking of these, how abominable it was to sread such rumors as were in the Boston papers last week. How worries every one must have been. I received your letter of the 19th on the 24th and one from Tom of the 23rd this morning. I went yesterday to see Mr Kemp He was in Washington but was expected back in the afternoon. I have found his stopping place & will pay him a visit soon & get my shirt. I wrote a long letter to Ned on the 20th and one to Jennie on the 23rd also one to Fan the week before. I have not yet received answers to anyone of them. Fan's letter together with a bundle of old letters I sent by Lieut. Atkinson of Lynn and Jennie's by Lieut Brooks of Haverhill. Both of these officers have been discharged on account of disability. I suspect all the above letters have been received before this Fan's and Ned's were written at a time when there was every possibility of an attack. Franklin's Division and Hooker's had both moved and ours was to follow the next morning, but all that night it poured in torrents & nothing has been done yet. Burnside has resigned & Hooker is now in command of the army. Sumner & Franklin have either resigned or been removed. Wilcox has been superseded in the command of the Army Corps by Sedgwick. Ferrero returned a few nights ago. Yesterday we went on picket in the midst of a pouring rain. Our brigade pickets from the Lacey House to Falmouth along the river. Hither I have been stationed near Falmouth but this time I managed to get in the Lacey House and had quite a comfortable time. The house is a fine specimen of an old Virginia mansion. It is situated on [—illegible—] a high bluff immediately opposite Fredericksburg and overlooking it. The view from it is magnificent. The building is of two stories with wings on each end. In one of these the ladies Aid Society have a room. Here I found two ladies Mrs John Harris & Mrs Parrish. The latter is the wife of Dr Parrish of Philadelphia. He was the physician who attended the college when I was in Burlington. I spent a couple of very pleasant hours talking over Burlington matters with Mrs Parrish. The Society contributed onions and apple sauce to my dinner. The officers had a room upstairs with a good fire. The furniture of the room consisted of a sofa and one chair. No carpet but none was needed. At night we spread our blankets in the floor and

turned in. We came back to camp through a driving snow storm. The snow is now six inches deep and the storm continues. It is horrible weather and it may last until April. It is pretty certain that there will be no better weather until then. Changes are as great as at home. Day before yesterday was as warm as a June day. I had no fire in my tent. Yesterday a warm rain which started the grass. The change in color was quite apparent. Today a cold driving snow storm. With all this mud, and mud such as must be seen to be appreciated. A winter campaign is simply an absurdity. Nothing can be done unless we have a fortnight of steady good weather. The letter Mrs. Mc Kew wrote to Lieut King was returned to me by an officer of the 20th Mass. Who said that there was no such person in the regiment. I sent the letter to Mrs Mc Kew When the order came to move I had to open Sullivan's box because there was no chance of taking it along or sending it to him. The contents I gave to those who tented out with him. Ned Rand sent me a cake of soap so I shall not want any more at present but an other time when I write for some or anything else which I want sent by mail please send it at once & call Mr Wakefield for the postage. Of course I don't want him to pay it. Tell him to take from my money the postage he had to pay on the two packages of mails he sent me. Postage makes no difference to me in comparison with the necessity of having the article. I have made an allotment of $100. A month which Mr Wakefield will draw from the mayor's office in Boston whenever we are paid off. This may be soon but I think not. The talk about paying the soldiers is a good deal of a hum. So far as I can learn they are [—not—] only paying them to the 1st of November last. It is said that the government fears that if the soldiers are paid in full thousands will desert. Pope was not promoted because promotions in our regiment generally go according to seniority & he was junior to those appointed. I am very glad that Capt. King is major. Joe I have not heard from yet. Don't be afraid of Joe wanting to reenlist. Not many
& socks if any
enlisted men will care to come out again. The mittens ^ had better be sent by express at once & my rubber coat may as well be put in the box. My valise and other things I can afford to risk at present. It is said that the plan of having three Grand Divisions in the army is to be given up. I need a pair of sock very much, & should like a pair by mail at once if you have not sent any lately. I spent two hours darning a hole as bid as half a dollar last week. What a treasure I will be when I come home. I can cook, darn stockings, sew, (?) build chimneys, log houses, sleep out doors & live on hard tack & coffee.
Love to all.

Yours affectionately Love

John

Newport News. Feb. 26. 1863

Dear Mother

I had begun to think that I was lost to sight and to memory too, for not a letter did I receive from the time of my return until last night when one came from you, one from Jennie and one from Jule. The last was a good long one with a postscript dated Feb. 11 saying the regiment would be off in two hours for parts unknown with 10 days rations, leaving tents standing. So we go, he has had an easy time of it so far while I have been in rough work, and now he is called on to do his share while I rest. It is perhaps premature to call it rest for there is no certainty of remaining here any length of time. At present it is delightful. Our camp is pleasantly situated just outside the fortifications at Newport News on a sandy plain. Everything in the eatable line can be procured at reasonable rates, and the we are close to the ocean which to me is the most agreeable feature of the whole. If we had such a beach at home it would be a fashionable watering place. Yesterday we had a review of the whole corps by Gen. Dix and Gen. Smith. The men appeared finely and the corps looked well. What is to be done with us no one knows yet. Some say we are going to Newbern others to Fortress Monroe, other to Suffolk. No one knows yet. Burnside is expected soon & when he comes he shall know more. One of my men died [—last n—} yesterday morning Feb 25th at a little after midnight in the regimental hospital of typhoid fever. His name is David K. Hall, of Needham. He was formerly a sergeant but was reduced to the ranks. During this month I have had three men who were absent in hospitals discharged for disability and one deceased here making my loss four. This is larger than the average. I have written to Mr Baker to have the box sent here at once. The box that was sent to Sergt. Cobb when we were at Pleasant Valley and which contained a pair of socks for me, arrived the other day. I have now an abundant supply of socks. I have been on Court Martial since I came back a good deal of time as President of the board and have a case to try in a few minutes. It seems quite natural to be at such work.

Capt Wm Gibson is going to Boston on a leave of absence today & I shall send this letter
by
[—to—] ^ him. Tell Jennie I shall write to her soon and give the same message to Mr Wakefield. I am well with the exceptions of a cold, but have not yet found my appetite. We had a tremendous storm last Saturday night & Sunday. Snow & rain in torrents. The time for the Ct. Martial has arrived & I must close. Love to all.

Your affectionate son

John

Camp Mass 35 Regt Mon
Sept 14th 63 Near Crab Orchard

Dear Captain

I have just received your letter of the twenty sixth of July and hasten at once to answer i. We are waiting here for our baggage train to come up so that we can proceed on our way to Knoxville Tenn. We are only allowed two baggage wagons to a regiment now and only shelter tents for line Officers the field and staff are allowed three tents. We have one small tent which we use for thee line Officers Gibson Ingell and myself Captain M[—y—]irich boards at the hotel with his wife. Lieut Meserve is acting adjutant and Lieut Tobey is acting R.Q.M. Lieut Berry is in command of Co H and Lieut Creasy is in command of Co F. We are so short of Officers that we have sent some of them home for conscripts. I am glad to hear that you are recovering your health though slowly you say you long to be with us and hope to be before long I can assure you I shall be most happy to welcome you back and I know the rest of the Officers will be glad to see you back also. You will to receive a heart felt welcome from the men of the company and regiment.

I showed your letter to the Major he was glad to know you are not jealous at his being promoted to the Majority of the regiment. You are now the second ranking captain of the regt.

In respect to the Clothing account of the company that is all right the clothing books were lost in transportation from Vicksburg to Cincinnati. And I have not as yet entered anything on the new company clothing book which I have just drawn. The books and papers were all left behind at Nicholasville for want of transportation

You say you were pleased at the selection of [—Bauer—] corporals and sergeants with the exception of Bauer. He has done very well since he has been a corporal and is certainly a fine soldier in camp. Rafferty has been reduced and is now reported as a deserter and is I believe in close confinement at Camp Nelson as a deserter he was left behind at Cairo on our return from the south this is the third time that he has stayed behind without leave and he will not get off this time as easily as he has before. There has been a number of changes in the company which you probably are acquainted with by this time through Titcomb and Smith and other channels

You wish to know what property I have belonging to you. I have one of your blankets the other one was lost in transportation from [?Suport S—-?] to Kentucky I have also belonging to you 1 pair blanket straps one camp stove one set casy's tactics the mess chest has long since gone by the board and there is only a few tin plates remaining of the old mess chest. I am messing now with Lt Meserve Capt Hudson also belongs to the mess you can mess with us when you come out if you wish. I have got the best darky that every entered the camp of the 35th Mass Regt I brought his from Jackson Miss he is a splendid cook.

You ask what became of the clothing that was in the company box that arrived at [?Suport S—?] after you left. it was delivered over to the Quartermaster by me by the order of Coln Carruth I got no receipt because Upton would give me none, and the Coln said it could be accounted for as lost for the want of transportation. I first gave the company all the clothing they wanted out of the box without charge and the rest was distributed among the other men of the regiment. It was all done by the order of the col. It makes no difference how regularly you send your certificates to the regiment if we do not receive them in time you have to be reported absent without leave until the medical certificates are received. though our not receiving them may be no fault of yours. I suppose Coln King will come back to the regiment now as Gen Sturgis has announced his staff and Lt Coln King's name is not on it. I believe Coln King got [?Mo—-kier?] at Gen Sturgis because he appointed [?ca—Irins?] on his staff of higher rank than Coln King; and Sturgis told him if he did not like it he [?mi——?] go back to his regiment_ the regiment is sadly reduced in numbers but there is but little sickness in the regiment Most of the sick are in convalescent camp at Camp Dennison We have less than thirty men with the company. John Hyde has returned to the company he is as tough as any of the men. Sergt Fiske and Calder with a number of the men are sick at Camp Dennison. Our Green Privates Downs, Albert Ellis, and one more man of the company "I do not remember who it is just now" are at Nicholasville doing provost duty. Sergeant Patch has been detailed as color sergt Sergeant White of company G, is sergt Major.

Envelope of letter to John from Albert Pope.

Your sincere friend

Albert A Pope.

PS

Samuel G Wright of our company is dead and Joseph R Manning is discharged

Harpers Ferry.
Sept. 3. 1865

My Dear Mary

I am writing on the assumption that on my arrival at Baltimore I shall find a letter from you awaiting me, and I Irish like, I will answer it. I wrote to Jennie on my arrival at Sharpsburg on Friday night. On Saturday morning I started off for the battle field where we were engaged. Crossing the bridge I went to this high hill where we [?s-ep-s?] posted a battery in the morning. From there I had a fine view of the country. While lying here on the morning of the battle the shells struck all around us, and we could see the rebels coming down toward the battle field; and the shells from the battery near us falling among them. This morning I identified the road down which they came. Passing down the hill I approached the bridge, and wondered how we ever contrived to cross. I then went up to the top of the hill, to which I thought then and think the more now that our batteries should have been moved after the bridge was taken. Then, down by the trees where Ned Davis was wounded to the bottom of the hill by the lane, where our regiment received its greatest loss. This was a spot of peculiar interest. The distance from the top of the hill to the bottom was very much shorter than I supposed. It was only a hundred of my paces. It seemed long enough while we were going down. I also measured the distance between the fence by the lane and the stone wall behind which the rebs were posted in a corn field to the left of the wall, towards which I direct the fire of my company. The position was this:

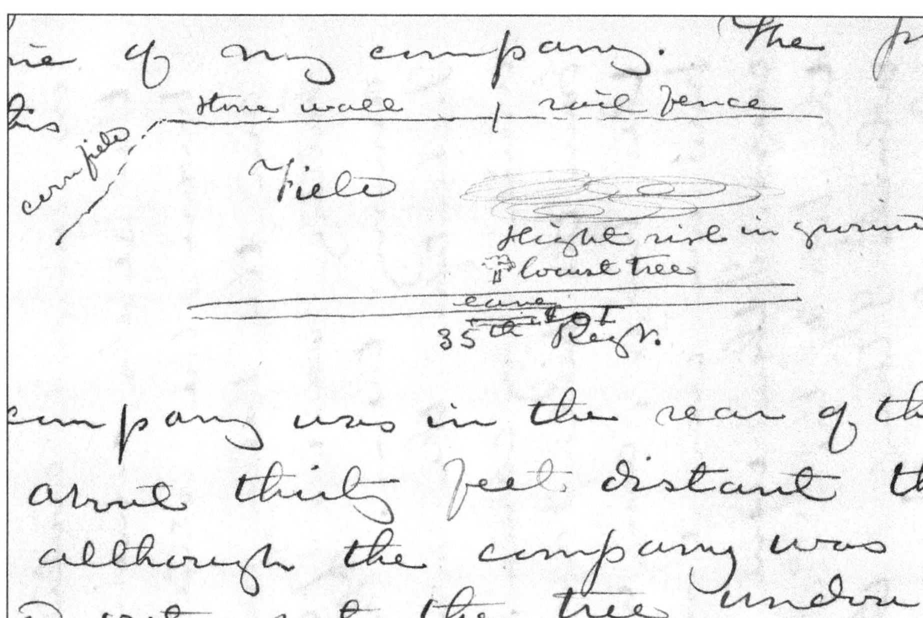

My company was in the rear of the locust tree about thirty feet distant there from and although the company was a good deal spent out yet the tree undoubtedly saved some lives, as the side towards the rebs was filled with balls. There was also a slight rise in the ground between us and the rebs served as a partial protection. On the extreme left where King's company was stationed there was no shelter whatever and they were nearest the cornfield where the fire was the hottest. From [—this—] a rail directly in front of where my company was I cut a bullet out of a peculiar and dangerous kind. It was cast like a minié but without a cavity. A solid plug of lead ran near through the ball longitudinally and projected a little beyond [—the end of—] the base of the ball. On this were thin rings of lead. The idea is that after the ball leaves the gun the rings will fly off, though they will go in the same general direction as the ball, and will thus cover more ground. [—There is—] It is also said that these rings are poisonous though this I think is doubtful. From this lane, the ground is level towards Sharpsburg for a few hundred yards and then rises very abruptly into a series of hills. The lane is about half or three quarters of a mile from Sharpsburg. Earlier in the day I was told in Sharpsburg that Burnside's skirmishers reached the town. In the afternoon I got a team and with my landlord as a guide went over this battle field where Hooker was engaged. I saw the famous Dunker church where the fighting was so fierce. Paint and whitewash have obliterated all traces of the fight, and the place looks the personification of peace. I also saw the cemetery described in the enclosed letter of Carleton's. Please put this [—let—] newspaper letter on my desk as I wish to preserve it. My guide was with the cavalry during the fight on Hooker's right flank, and the next morning went over the ground. This was interesting ride, and if the carping critics of the North should see the positions from which they [—would—] rebels were driven and the line which they held in the morning and the one they held at night, they would conclude that the victory was somewhat more than a drawn one. I also learned what I did not know before that the rebs did not have all their troops engaged in this fight, but had a large force in reserve. Sharpsburg [—has—] sent three Union companies to the war, and a dozen or more were in the rebel army. Some of these men were in the fight, and they say that Hills corps, the one that took Harper's Ferry were not engaged during the morning and but a few of them in the afternoon. They also said that the rebs had quite a large force at Sheppardstown which is back of Sharpsburg across the river. Indeed some of the Sharpsburg men say they can not see how Lee could make such a fight as so great a part of his army was the other side of the river. I think likely Lee's notion was to keep a force in reserve with the hope that McClellan would put all his men in and then when our men were all tired out, to put in his fresh troops and to win the day. This morning I started off about eight o'clock and drove by our old camping grounds near the mouth of the Antietam creek, through Crampton's gap, now called McClellans where Franklin had his fight, Sept. 14. 1862. Thence I drove through Pleasant Valley where we were encamped in Oct. 1862, and arrived here, hot tired and hungry. Soon after I arrived it commenced to rain, and we had some refreshing showers. The hotel is beautifully situated close to the railroad bridge with the Potomac on one side and the Shenandoah on the other. It is however miserably kept, and although larger and dearer than the other hotels in this region, is more dirty. The general style since leaving Frederick has been one sheet. Here they have two, but they are not so inviting as one clean one. The people here are far beyond the age and seem to have no idea as to how a hotel should be kept. Tomorrow morning I propose visiting Maryland heights and in the afternoon, go to Baltimore and thence on Tuesday to Washington and Fredericksburg. I may not write again until I reach Fredericksburg or Richmond. I hope to find some letters from home in Baltimore on Monday. Love to all

Your affectionate brother
John

Julius Lathrop

JULIUS LATHROP LETTERS

On board the Steamer Admiral at Sea Dec. 11/6

[Note: There was a "0" added below the "6". The ink color and the handwriting appear different than Julius'.]

My Dear Mother

I presume you want to know all about our journey, reception and embarkation, so now I will endeavor to give you every particular. As John and Joe told you of our departure it will be unnecessary for me to write about that, so I will commence at the time the train started. Mary Gardner and John were the last familiar faces I saw, as they were beyond all others and as we passed them I waved my handkerchief but do not think that John saw me though Mary did. All along on the route people cheered us and whenever we stopped the band would commence playing. Our company occupied the two cars just in from of the one the field officers and Band were in and I happened to be in the end nearest them, had the best chance to hear. At Providence the train stopped for a short while so that we could get something to eat, that is those who wished to pay for what they took. The band in the meantime played are great many tunes and the house tops being covered with ladies waving handkerchiefs, we enjoyed the stop very much.

After leaving Providence nothing special occurred till we arrived in Groton about 7 o'clock where we all embarked on board the Vanderbuilt for New York. I thought we never should leave but after waiting till I got tired hearing some of our men had run guard and were trying to get rum or something else to drink Sergt Perkins Sergt Turner (a jolly good fellow) and myself started up [?] after them but by the time we got there the men were not, so back we went and just in season of the starting which took place a little after ten. As the fog was very thick we had to wait some time at anchor about ten miles below New York, till the air was clear enough to see, when we proceeded on our voyage. At 12 o'clock the boat landed us at the foot of East 23rd St where the line was formed and we marched on preceeded by the Sons of Massachusetts and Dodsworths band, through 5th Av. To 14th St where we turned up Broadway and continued to the Park

I must confess that I was a little disappointed at our reception as I had expected much more enthusiasm, but every one said that no Regt had such a good reception as we had, and all along the route I heard everybody praising our marching tc. Being on the right of the 2nd platoon I had a good chance to see all the young ladies and hear what they said of us, which I can assure you was very flattering. We must have marched near five miles and as the day was very warm we were pretty well used up by the time we reached the Park There a good dinner awaited us & to which we did justice. About 6 o'clock the left wing, consisting of our

Co and four others, was formed in line and preceeded by Dodsworths band started for the Steamer Admiral which was to carry half the Regt. As we left the Park the band struck up "Viva
there
l'America" which in a minute put me in mind of home as the last time I was ^ I had Louise Adams sing it for me. There were plenty of cheers and good byes for us then as we marched up Broadway and down Canal St to the boat, and we fully deserved all the applause as our marching and wheeling was very fine. Our band is on the other Steamer behind us and in the evening while we were at the wharf they played for a long time. We all think that ours is a much better band than Dodsworths as they certainly played much better.

At 7:30 this (Wednesday) morning we started for Annapolis by Steamer much to our regret as we all expected to go by cars, but as we could not help ourselves, we had to grin and bear it.

The morning was very thick with every signs of a storm, which overtook us before we had passed the Narrows and continued till the middle of the afternoon. Just as we started we saw a splendid double rainbow, although it was not raining then so we all hailed it as a good omen and has thus far proved true for everyone who has ever been here before, [open space] agree that the sea has been very smooth, still for all that a great many were sea-sick and are now.

Thursday morning Dec 12th

Last night when we got opposite the Delaware Bay the water became very rough and of course we were all pitched about pretty lively but as yet I haven't felt the slightest degree sick and don't think I shall now as we are at the present moment (9 o'clock) in sight of land again which I presume is Cape Henry although I am not sure. So far we have made a very fine passage and by nine o'clock to night we shall without doubt be at our journey's end.

In New York I saw Ned Fisher who goes out as clerk for Gen Burnside so you see Dedham will be represented pretty well in this expedition.

You need not expect me to write very well while I am on the Steamer as we are "reeling to and fro like a drunken man" so that it is hard work to stand or even sit still.

I shall send home $5 to pay my boathouse bill which I think amounts to $4. The balance you can keep till I give further directions as I believe I shall have a few photographs taken, having promised one to Mrs Rodman and one to Miss Parsons for the present she made me of a writing case. However the next time I will decide.

Just now the mate of the Steamer told me that in the night we ran into a whale, whatever it was we struck it pretty hard as I felt the shock severely.

At present I am sitting in the upper salon which is pretty well filled with the non-
semi
commissioned officers ^ engaged in writing others in playing cards, so you see it is almost impossible for me not to make mistakes in spelling, in fact it is hard work to write at all for the noise and confusion.

I think now that under the circumstances you have got a pretty long letter from me, still when we get settled at Annapolis I will write more about our voyage.

Saturday Morning Dec 14th

Well here we are at last in camp and we are as it should be. Thursday while we were coming up the Chesapeake Bay we saw thousands of ducks all round the Steamer in every direction and of course all who had pistols were practicing, but on account of the rough state of the water but very few were killed. Yesterday we that is the left wing, landed and marched out some two miles from Annapolis to our camp, escorted by the 23rd Mass Regt who have pitched their tents in a grove of pines on the spot formally occupied by Gen Washington and remains of earthworks are still to be seen. I enclose some holly leaves which I picked on the top of the remains of the breastworks.

Yesterday afternoon I went off foraging and obtained a lot of corn to pop besides any quantity of persimmons; enclosed you will find a little of the former taken from a seceshes farm. Today I started for a tramp on the woods and am now in company with four others all engaged in writing home. We are away from camp in the woods and this part of my letter is being written on the top of my cap for a desk.

All the men are now getting holly and evergreens to decorate the tents and place round them, after the fashion of the 23rd.

Coons and quails are very abundant here and can be seen most anytime if care is taken to look after them. I only wish I had my gun here.

Camp Foster Sunday Afternoon Dec 15

This morning the right wing made its appearance after meeting several misfortunes on the way. Col Stevenson and the first five Companies embarked on board the Steamer Eastern Queen expecting to reach Annapolis in advance of our boat, but instead of that we took the lead and kept it, arriving several hours in advance besides running all Thursday at half speed. In addition to being beaten the Steamer ran aground a few miles below the city, where she now lies, and yesterday all the troops had to be taken off on small boats. Last night they camped in the Naval Academy and marched up here today.

I enclose a photograph which I want Joe to carry down to the office and give it to Somes as I promised him one some time ago and forgot to leave it. I also send you my certificate of enlistment so that if anything should ever happen to me you can draw my bounty money and land warrant. Take good care of it.

Every day we have been here I have spent mostly in getting persimmons and running over the country. The weather here is glorious in the daytime though cold at night. Now I am sitting in the woods about a mile from camp writing and while I presume you are shivering with cold I am roasting in the sun.

My health is very good and so far our being here seems more like a picnic than anything else as we go where we want to though I presume before long drilling will commence.

Give my love to all and write soon directing your letter to me at Annapolis Md.

Your affectionate son

Julius

P S Night before last a secesher was killed while attempting to pass the pickets a short distance from here.

In a day or two I will write again

Yours in haste J

Camp Foster near Annapolis Dec. 26th/61

My Fan

I received your letter and two papers Tuesday afternoon. You can be assured that they were very welcome as we don't very often get any Northern journals of news and articles of that sort are very acceptable.

Yesterday did not seem any much like Christmas except that the band marched round through all our streets playing and no drill took place, beyond that, everything went on as usual.

We had a visitor from the 18th Regt. Sergt. Pond of the Dedham Co. He reports all the boys are well and enjoying themselves. The Regt just received a new uniform, tents and all other articles necessary for a soldier from France so that now they are splendidly furnished far better than we are. Their uniform is that of the French Chasseur and is very gay. There is some talk of their joining our expedition although I don't believe it, yet I hope it will prove true as they are a very good regiment, and then it would be better for me to have some Dedham boys along with us. Those papers I received from you, have gone on to the boys ^ from Dedham and I have no doubt will be very acceptable.

We can't fund out any thing relative to our expedition as it is kept very secret, and I don't believe we shall start for some weeks, yet we may go in a very few days.

Christmas eve the band played for some hours and you can't imagine how good it was. They are the best players in this part of the world and of course we have a great many visitors from the City (plenty of ladies) and from the other camps so it makes it very pleasant for us.

That night Lieut Bell's servant took a revolver without leave and while playing with it accidentally shot the Quarter Master's contraband wounding him so badly that he died yesterday morning. His makes the second death which has occurred since we came out here. The first was a man in our Company who died of congestion of the lungs. He was out on dress parade one morning and in less than twenty six hours after, was dead. We had a military funeral and after that the body was sent home to Cambridge. In the morning we usually have battalion or brigade drill and on the afternoon our company learns the skirmish drill. This morning the right wing commenced target practice while the left wing acts as skirmishers. Tomorrow our turn will come for shooting.

I want you to tell Joe to get for me three boxes of cartridges such as I enclose. He can take a dollar which he has of mine to pay for one box and I will send home some money as soon as we are paid off, which will probably be in a few days. Have them sent as soon as an opportunity offers. I am not in much hurry for them, but shall have occasion to use the little jokers when we get in a battle, as I have to reserve my fire till the colors are attacked, but can use my revolver in action.

I haven't been in the old state house yet, but probably shall in a few days. Capt. Stackpole is now in the city as he isn't very well and the Doctor thought best for him to keep quiet for a short time. He has a bad cold and is a little feverish, yet he is not unwell and without doubt will be with us before long. Last night his servant brought us word that he was better.

The idea of Annapolis being called a City. It is perfectly absurd as I don't believe, taking out the soldiers and darkies, there are as many inhabitants as we have in Dedham.

At last we are getting a little chilly weather, as for the last few days we have had to drill with overcoats on.

It is a grand thing for me to have obtained my situation on the colorguard, as here I am sitting in my warm tent writing while the rest of the boys are out drilling. Still whenever we have skirmish drill I usually turn out.

As it is too late to wish you all a happy Christmas, I hope you will have a jolly good New Year. Kiss little Johnny and Neddy for me and tell them I often think of my little nephews.

How I should have enjoyed myself, if I had been home yesterday. I could not help thinking what a good time you are having eating dinner, while I was devouring mine consisting of roast beef, hardbread and hominy.

Why don't some of you write me how you all are at home? I don't know whether you are sick or well, so now the next time give me such information as you can, as I shall fell much better to know if anyone is unwell.

There is nothing here to write about, so you musn't expect a long letter from me, but when we arrive at our future destination then you will get a good one.

Send me a Post now and then as we don't get any too much reading here. Now and then a paper boy comes along so we get the news little by little.

Give my love to all and write soon to your affectionate

brother Julius

Camp Foster near Annapolis Dec. 26th/61
My Fan

I received your letter and two papers Tuesday afternoon. You can be assured that they were very welcome as we don't very often get any Northern journals of news and articles of that sort are very acceptable.

Yesterday did not seem any much like Christmas except that the band marched round through all our streets playing and no drill took place, beyond that, everything went on as usual.

We had a visitor from the 18th Regt. Sergt. Pond of the Dedham Co. He reports all the boys are well and enjoying themselves. The Regt just received a new uniform, tents and all other articles necessary for a soldier from France so that now they are splendidly furnished far better than we are. Their uniform is that of the French Chasseur and is very gay. There is some talk of their joining our expedition although I don't believe it, yet I hope it will prove true as they are a very good regiment, and then it would be better for me to have some Dedham boys

from Dedham

along with us. Those papers I received from you, have gone on to the boys ^ and I have no doubt will be very acceptable.

We can't fund out any thing relative to our expedition as it is kept very secret, and I don't believe we shall start for some weeks, yet we may go in a very few days.

Christmas eve the band played for some hours and you can't imagine how good it was. They are the best players in this part of the world and of course we have a great many visitors from the City (plenty of ladies) and from the other camps so it makes it very pleasant for us.

That night Lieut Bell's servant took a revolver without leave and while playing with it accidentally shot the Quarter Master's contraband wounding him so badly that he died yesterday morning. His makes the second death which has occurred since we came out here. The first was a man in our Company who died of congestion of the lungs. He was out on dress parade one morning and in less than twenty six hours after, was dead. We had a military funeral and after that the body was sent home to Cambridge. In the morning we usually have battalion or brigade drill and on the afternoon our company learns the skirmish drill. This morning the right wing commenced target practice while the left wing acts as skirmishers. Tomorrow our turn will come for shooting.

I want you to tell Joe to get for me three boxes of cartridges such as I enclose. He can take a dollar which he has of mine to pay for one box and I will send home some money as soon as we are paid off, which will probably be in a few days. Have them sent as soon as an opportunity offers. I am not in much hurry for them, but shall have occasion to use the little jokers when we get in a battle, as I have to reserve my fire till the colors are attacked, but can use my revolver in action.

I haven't been in the old state house yet, but probably shall in a few days. Capt. Stackpole is now in the city as he isn't very well and the Doctor thought best for him to keep quiet for a short time. He has a bad cold and is a little feverish, yet he is not unwell and without doubt will be with us before long. Last night his servant brought us word that he was better.

The idea of Annapolis being called a City. It is perfectly absurd as I don't believe, taking out the soldiers and darkies, there are as many inhabitants as we have in Dedham.

At last we are getting a little chilly weather, as for the last few days we have had to drill with overcoats on.

It is a grand thing for me to have obtained my situation on the colorguard, as here I am sitting in my warm tent writing while the rest of the boys are out drilling. Still whenever we have skirmish drill I usually turn out.

As it is too late to wish you all a happy Christmas, I hope you will have a jolly good New Year. Kiss little Johnny and Neddy for me and tell them I often think of my little nephews

How I should have enjoyed myself, if I had been home yesterday. I could not help thinking what a good time you are having eating dinner, while I was devouring mine consisting of roast beef, hardbread and hominy.

Why don't some of you write me how you all are at home? I don't know whether you are sick or well, so now the next time give me such information as you can, as I shall fell much better to know if anyone is unwell.

There is nothing here to write about, so you musn't expect a long letter from me, but when we arrive at our future destination then you will get a good one.

Send me a Post now and then as we don't get any too much reading here. Now and then a paper boy comes along so we get the news little by little.

Give my love to all and write soon to your affectionate

brother Julius

Camp Foster Roanoke Island NC Feb 13th/62

Dear Mother

A few days ago I gave you a hurried account of the fight and its results, and intended writing further as soon as we got back on the boat, where I could have paper & ink, but there being no prospect of our leaving here just now, will endeavor to write more about the part our regiment took in the attack. In the first place a short time after the gunboats commenced the action by firing at the rebel fleet under command of Com. Lynch and the batteries on shore, the signal was given for the transports to move up nearer the Island as we were then about twelve miles off. Our regiment was to have been the first one to land and the colors were to go in the first boat, but scarcely had we started when the steamer ran aground so we had the mortification of seeing all the other regiments pass by us, while we were left lamenting. There we remained watching the shells bursting around the boats. It was not long before the shells fired from our side set the rebel barracks on fire thus adding to the already magnificent scene. That night all the troops were landed and slept on the beach in the rain which commenced early in the evening. The shelling of the woods & batteries lasted just seven hours from 11 til 6 and the latter part of the time, as it grew dark the bombardment was far better than any fireworks as we could see the flash of the guns & the bursting shells.

Early the next morning the rebels commenced firing on our pickets and soon two brigades were pushed forwards driving the rebels to a masked battery, which I will now endeavor to describe. Imagine a swamp beside which Wigwam swamp would be open ground, and one which the secesh thought no one could ever penetrate, as it is deep mud & water with trees , thorns, briars and bushes so thick that one cant see but a few feet on any side, and only one road approaching the embankment, ^ which were three big guns. The trees & bushes for about 300 yards in front were cut down so that they had a glorious chance to rake our troops with the rifles and cannons. In front of the battery was a wide deep ditch, so that storming it in front was impossible. Our troops were marched up to the open space & then deployed as skirmishers, each one firing when he had a chance, while the rebels swept the woods with bullets, shells, grape & canister and as their men were protected by the breastworks but few were killed. If we could have brought up a few cannon & had a solid place to have used them, the number of their dead would have been largely increased. As it was the larger part of their loss, was in their reserve stationed about half a mile behind the battery. The "Wheelbarrow". We came along side of our steamer & took us off so that we landed a little after ten and while the firing was still going on. Our line was formed in front of the buildings used for the hospital & while there had our first view of the wounded who were being brought down very fast. As soon as Gen Burnside came up, the order was given to start forward and away we went all ready for the fight. A great many of the wounded cheered us on, especially our brass howitzer which was dragged through the mud at the head of our line. Just before we reached the scene of action three rousing cheers were heard and at the same time a tremendous volley. After we found it was the Zouaves & our regiments, who had crawled through the swamp and flanked the enemy at the same time under cover of rifleballs, charged the battery at the point of the bayonet. We pushed on & passed by the Lt Col of the 53rd NY who had gone into the fight as a private and was killed in front of the fortification. I saw another poor fellow who was shot through the head with a grape shot, & was still alive although his brains were running out of the wound. After all I was not excited by the sight of the wounded & dead although plenty of both were in sight, from one with a small hole in his head or body to others cut nearly in two by cannon balls. I saw all sights of wounds made by balls, round shot, shells, grape & canister. When we started from the beach all expected that we should be "in at the death," but it was reserved for us to the "biggest thing", for we were order to push on & support Gen Reno who was pursuing the enemy. So on we went double quick through mud & water until we had overtaken the troops, those were soon passed, so that we at last were ahead of the crowd & close behind the rebels, who we thought had gone into another battery. Nearly all the way Gen Foster rode beside our colors, so that I was within a few feet of him or at his horses side. Every minute I expected to see him drop, as enemys' sharpshooters could have easily picked him off, & we afterwards learned that two had remained in the bushes for a long time waiting for a shot, & were taken prisoners before they had a chance. A rebel major came down from their camp & proposed to surrender. Ten minutes was given for our Major to go up & get back, at the end of that time not having made his appearance we were ordered on doublequick. Just as we turned up the road near here, about a Company of "rebs" came out of a house & gave up their arms. Part of our regt was detailed to guard them & go out scouting so that only two companies marched in before over 2000 troops most of whom had arms in their hands. Well to make a long story short, the rebels themselves say that if they had had fifteen minutes longer they would have made another stand, so you see it was all for the best that we were not in the fight, for we had been, we could not have pushed so fast so that there would certainly have been another big battle. We get the credit for this as they surrendered to our Col & their colors have been sent to Gov Andrew. You will probably see any quanitity of butcher knives in Boston before long taken from the rebel troops, with which they said they intended to murder our wounded. I picked up a few letters here & shall send them on before long. I might have had a good sword, but could not carry it.

The weather now is glorious being very warm & like what we have in June.

I must close now as a mail will leave in a few minutes. Let anyone at home read this who want to, as I can't write to every one. Excuse dirt, but I must tell you that this is secesh paper of course can't be clean.

We have captured about 2500 prisoners without the loss of a man of our regt. Tell Joe to say to Mary Gardner that her friend Frank Oakes with a few others captured tow rebel officers & several privates. I am very well & in good spirits. Write soon & send some papers especially those having pictures of our fleet &c. One of Frank Leslie's Artists is here & was sketching this camp. If you see a picture of it send it on, as I took the "Color guard's tent & five"

Love to all & all my friends Yours in haste

Julius

Camp Foster Roanoke Island NC
Feb 16th/62

Dear Mollie

I received your letter a few days after writing the last time and having plenty to tell you, will endeavor to scribble off a few lines about our victory, although without doubt you know more about it now than I do still there may be some little things not published in the papers. First let me tell you that a letter from home or ones friends is always very acceptable, and our last mail on the boat brought one from Fran, one from Joe & yours. A few days ago I received one from John & Joe with a short note written by Mother. What fun (?) it must be at home judging by the accounts. I hear of snow and ice, while we don't know of such things out here, on the contrary the birds wake us up every morning, and the days are very warm, just such weather as we have in May. The last time I was outside the lines, two of us went in swimming. Well to commence. It was intended that our regiment should be the first one to land and as the colors were going in the first boat, I should have had the extreme felicity of being one of those in advance but the saying of "the first shall be last" proved true in our case for when the signal was given for the transports to go into line and move up the narrow channel in the rear of the gunboats, our steamer the "Guide" ran aground. (I nearly forgot to tell you that our Co went on that Starr a few days before we left Hatteras.) Here we remained hard and fast and had the mortification of seeing all the other boats steam slowly past us and proceed to the naval fleet which was engaging the enemie's boats and batteries. A little after 11 o'clock a few guns were fired by the rebel squadron at ours, and a short time after the action became general our boats soon driving the rebels out of reach, then they opened on the batteries. It was a sight never to be forgotten and one which I have always wanted to see. The shot & shell flew fast and we could see them strike the water on burst in the air. Those from the batteries fell short but ours went over their fortifications and soon a dense cloud of smoke arose, caused by the burning of their barracks. Gradually their fire slackened although ours still went on, the shells sometime going over a mile into the woods, and even bursting inside of the masked battery taken the next day a mile and a half from the beach. Our boats after awhile silenced a battery so that at the first attempt our 3000 troops were landed safely, and soon the remained were on shore. A small cornfield near the beach was their camping ground and large fires lighted to keep the men warm for a heavy rain storm came up, and you can imagine how very pleasant it was for the soldiers, most of them without blankets and wet through. Very little sleep was had that night as it was almost impossible. On account of the cold and rain together with the occasional firing by the rebels on our pickets. That night Gen Foster went out scouting and discovered a masked battery with three guns in the midst of a swamp & all the trees and bushes in front for several hundred yards cleared out so that no cover could be had, for a storming party. Very early in the morning the rebels opened fire on our men and soon the troops were engaged in what the secechers term "bush whacking" firing on each other when ever a chance was had. It was not long before they retreated into the battery, thereby thinking we should go into a trap. , but our Generals knew better having had experience at Bull Run, for instead of approaching it in front our men were deployed as skirmishers and advanced into the swamp which I will now describe.

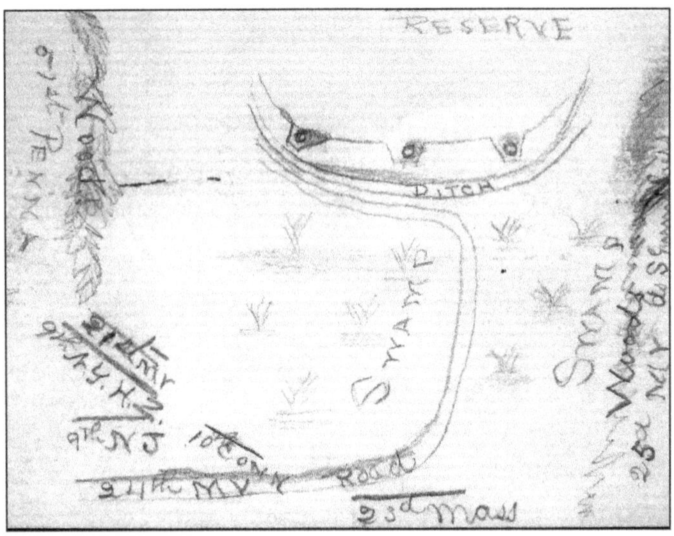

Map drawn by Julius of the Confederate masked battery, shown in the upper right with three guns facing the Union troops. Directly in front of the battery is a swamp with a road bisecting it. Union forces arrayed facing the battery include, from left to right, 51st Penna, 21st MV, 9th N.Y.H.Z., 9th NJ, 24th MV, 10th Conn, 23rd Mass, and 25th NY.

The mud is fearfully deep with water in some places above our knees, then the heavy underbrush, briars, and prickly thorns and a species of laurel, with trees made the place almost impassable. In fact only two or three weeks Gen Beauregard was out here and pronounced that the place could not be taken by 60,000 men. The Southerners call us "Mudsills" and we fully deserve the name, for the way the men went through the mud and water was a caution to all having secesh proclivities. Some of the prisoners have told me they never supposed we could get through that swamp. Just as soon as our troops could get along they commenced firing at the rebels whenever a head showed itself over the ramparts. The fort itself was of turf in the shape of a crescent in front of which was a deep ditch. At the end of this letter I will give you a sketch of it to show a little more fully the place where the "rebs" expected to wipe out the Mass Boys. Their guns sent grape, canister & shell crashing thro the bushes and knocking our poor fellows over lively. All this time while the fighting was going on, we were being taken on a smaller steamer & landed in the mud where the troops came on shore the night before, and our line was formed in front of some buildings then filled with wounded. It was the greatest trial that green troops could be subjected to seeing the wounded and the dying brought in front in long strings, sometimes twenty or more at once, but although severe, we stood the test well and I did not see a man turn pale although we could not tell how soon it would be before some of us would be carried back on litters. We halted for a few minutes for Gen Burnside to come up and I meanwhile took advantage of the time to examine my valor to see if the charges were all right, and then the Gen told us he should go up with us and wanted us to do our duty. On we tramped double quick and just as we reached the open space, [indecipherable word crossed out] rousing cheers were given, [crossed out letter or character] with a tremendous yell Hawkin's Zouaves (the 9th NY) rushed in at the point of the bayonet. The rebels seeing that long line of steel & the redcaps almost on them, broke & ran, those remaining were soon shot or bayonetted. The Capt of one of the guns stood beside it all ready to fire, having the lanyard in his hand. The first ZouZou on the top fired at him shooting him through the body making the most awful wound I saw and then the fellow drove his bayonet into his head. If his gun had missed fire, a large number of our men would have been killed as the cannon had just been loaded with grape. I saw one poor fellow carried by us shot through the head with a grape shot, and although his brains were running out, yet he was alive. The first dead one I saw was the Lt Col of the 53 NY who had gone into the action as a private, his regiment having been at Fortress Monroe. I saw all sorts of wounds from a rifle shot to a cannon shot. As I told you we were just too late to take part in the fight so we pushed on through the battery where I was this capt, whose death I have described, beside several other dead rebels and halted in the rear, for a few minutes to rest. While we were marching just in front of the battery who should I see but Chip Hopkinson who is a Capt in the NJ 9th and has been acting as Major since Col Allen's death. Tom Middleton is also a capt in the same regiment and I make them visits every day as they are close by us. I nearly forgot to tell you that the Zouaves fired on some of our regiments by mistake killing & wounding quite a number. We did not halt long as Gen Foster ordered us on to support Gen Reno who was then chasing the "rebs". We double quicked it for several miles until we had overtaken the advance troops who were resting for a short time. We passed them, as one of the 21st says, "like a train of cars" and Gen Foster being just beside our colors I was all the time either in front of him, or at his horse's side and every moment expected to see him drop from his horse as the woods were full of rebels and he was a splendid mark being the only one on horseback. We halted again a few miles further on to wait until reinforcements could come up when a rebel major came down with a flag of truce, to propose terms of surrender, Gen Foster told him "unconditionally" and sent our Major back with him. After waiting a few minutes Gen F said something was wrong and it was "forward boys double quick". Just as we turned up the path leading to this camp we saw a crowd of the enemy coming out of a house with a flag of truce and some of our companies were detailed to guard them & others to go out skirmishing, so that we marched in here with only the Color Co & Col these were the only men for over ten minutes, by which time others came dropping in. Well your beloved brother with the colors (how the seceshers have cursed our flag when they saw the "24th Mass") formed in line with about 2500 Armed rebels in front of us. I expected nothing else but a volley, although a white rag was flying on the Col's quarters, and had my eye on a group of officers, so that if I was not killed at first, determined to make one or more "go under" if any attempt had been made to shoot us. The enemy themselves said if they had known we were the only ones near they would have shown fight. Talk about Bull Run why that run was nothing compared with this, as they threw away guns & everything they had, coats, shirts equipments, knapsacks and canteens. You can judge how they went, when we, a whole regiment which had not been in the fight & consequently not tired, could not catch those who had been in action for several hours, although we did our best. Our march was a tought one through mud& water for over eight miles, so you see the 24th had their share of the action, altho' "too late to be in at the death", yet the rebels say if they had had a few months longer, they would have shown fight, and already one regiment was drawn up in line of battle. The 24th took their colors & have been sent on to Boston, and they get all the laurels much to the disgust of the other regts, as all the rebels surrendered to our Col. We are all to have inscribed on our Colors "Roanoke Island Feb 8th 1862."

The rebels taken here are the flower of the Southern Army being Wise's Legion. Old Wise himself was sick on the main land & escaped being taken, although his son was wounded so that he died Last Sunday morning. One of the Companies here is the Richmond Light Infantry Blues, all first class fellows. They are the ones who entertained the Boston Knight Templars in Richmond two years ago & our band being with them, made many friends. The meeting this time was rather different than the first. When our music struck up, the rebels were rather astonished as most of them had never heard any band worth listening to before. They rejoice in many names such as Sulphur Wild Cats, Charleston Sharp Shooters, McCollough Rangers, Kanawha Boys & c We took in all about 3500 prisoners two hundred and fifty of them officers, three Cols. I do hope now that Cols Cocoran & Lee will be exchanged as all of them will be sent home in a few days, they taking the oath not to fight again until an equal number of ours are sent north. This is the "biggest thing that has yet taken place, so many prisoners surrendering to the 24th without firing a gun. Aren't you proud of the 24th doings?

Do you know Benj H Mann who graduated at Union College last Summer? He is now writing beside me and begs to be remembered to all the girls in S. He knows Hic and in fact all the good folks in the City. He is now Hospital Steward in our regiment and we have made up our minds to visit your town <u>just</u> <u>as</u> <u>soon</u> <u>as</u> <u>we</u> <u>get</u> <u>home.</u> Last night I saw in the papers of Judge Wright's death and am very sorry to hear of it. I little thought that when I was in S, it would be the only time I should see him. I have a button which I cut from a secesh Majors coat which I will send to you before long. I could have had the Col's sword, but couldn't very well carry it, although am very sorry now that I didn't take it. I confiscated his pipe with the initials W.J.P. for WJ Price Col 10th N.C.V and a piece of the Majors sash which I now send you. Most of the boys got revolvers or big bowie knives, the latter the secesh say were intended to be used to cut our throats with, when we were wounded, and I believe they would do it. I for one after this will never show any quarter to such fellows, and if I am wounded shall put my trust in my revolver, so that if one of the "devils" tries that cut throat game, he will find that two can play at it. I have no doubt, but that before long every store in the City will have these knives on exhibition.

I shall get one to carry home if possible.

Please send this letter home for the[illegible] to read as it has more particulars than either of the two I have yet written, although this is a little larger edition, as before I had to write on secesh paper and that was very scarce, but now having got my knapsack, things are more to my taste, as I think pen & ink are preferable to a poor pencil.

I wish you could have seen some of the letters we confiscated here, it would make a cat laugh. One of them, a woman writes to her lover that if he will come home she will "kil a fride chiken, churn and crack a egge &fri its branes." this is verbatim [cross out et] literatim. Enclosed you will find a love letter from a romantic schoolgirl to a longhaired secesh. I want this sent home & tell John to get it printed in the Dedham Gazette for a copy for the youthful minds in that town. It is very rich, though the paper is very dirty, probably her tear marks. In a day or two I shall send one or two more letters to John, so that you can see what kind of people we have to deal with.

Our color guard, Dr. Mann and two of the band have got a jolly big tent, where we quarter and you can bet your life we live well, doughnuts, fried cakes, rice cakes & such "stuff" are the results of our cooking. We now have the floor all boarded over and a small stove goes "bully" Yesterday I have the first fresh meat I have tasted since the 6th [Ian?]. Although many pigs & cattle have been killed since our landing here yet it hasn't been my luck till yesterday to get any. <u>Now</u> we have two rations of meat a day, so that soon Kichard will be himself again.

I am going over to see Chip Hopkinson & Tom Middleton in a few minutes & will finish this letter tomorrow. I think the former will be the Lt Col of the regt as all the officers have sent on a petition to that effect.

Thursday night we received a mail & I got a letter from John & one form Joe also two papers, one of them containing an count of our expedition. It was very good and very true in all particulars. One of our Color Guard rec'd 29 papers besides several letters. We have just heard that all bands are to be discharged, if it is true we shall lose the best band which has yet gone out to the war. I rather think there will be a big row if the bill has passed. There is nothing so cheering to the soldiers as music, especially after a long march and if there isn't a little grumbling, then I am much mistaken.

Monday morning

I spent yesterday afternoon with Chip Hopkinson and Tom Middleton and had a very pleasant time. The both beg to be remembered to you & all at home. As the mail does not go out until Wednesday I may write a few lines home, but you had better send this on as soon as you have read it. Yesterday telegraph poles & wires went up on the Island connecting with a cable to Hatteras Inlet so that we may be able to hear a few more lies. The first prisoner I saw here told me that one would hear more false reports on this Island in one day than anywhere else in a year, and I believe he spoke the truth.

We call ourselves in this tent Col Stevenson's happy family as we do get along as the secesh say, "right smart, I reckon."

This afternoon I am going to the Hospital with Dr. Mann to see the wounded numbering about seventy five. The most cheerful of them all is a poor fellow of the 9th NJ who has both legs shot off below his knees, although badly wounded, still be will recover.

Before many days, I hope we shall leave this place & go further on where the 24th may win more laurels. Gen Foster wanted our regt to go home with the [illegible] over, but Col S told him that he wanted to show his men in battle. The Gen replied he was perfectly satisfied with it.

Now Mollie dear don't worry about me as there isn't much danger of my being hit as our experience shows us that the "rebs" won't give us a fair fight & in future all the masked batteries will be taken at the point of the bayonet, that being the safest way for our men, & the worst for the enemy.

Give my love to all my friends also to Hic. Tell him that a letter would be very acceptable. I am in tip top spirits & health. Henry Birney is also well & is now in the tent near me.

Good by now & write soon to
Your affectionate brother
Julius

Camp Foster Roanoke Island NC
Feb 19th/62

Dear John

I received all the letters from home last Thursday & now having a little time before the mail closes will send you a few lines. I am very much obliged to you for the papers containing the account of our Expedition and can assure you they are very true in every respect. I know of the drowning of Col Allen and thought I mentioned it in some of my letters. The vessel he was on laid very near ours at the time of the accident, although we didn't hear of it for several days.

I have written a letter to Mollie containing the full particulars of our part of the victory and requested her to send [oh] on as soon as she got through with it, together with a secesh love letter which I think you had better have published in the Dedham Gazette, as it will look better than the "taxpayers" I send you several other letters which I captured, showing a specimen of the NC writers. Our of over two hundred prisoners there were only sixteen who
of men
could sign their names and this is a very fair example of the style ^ we have to deal with. One of the letters you can give Ned Sumner, either one you please, and tell him I should like to see a little of his hand writing as a letter would be very acceptable to me just now. I also enclose you a button which I cut off a rebel Majors coat sleeve, keep it with the other rebel articles.

You talk of snow, ice and sleighing, things we don't know out here in the North Carolina swamps, while I can say we hear the birds singing every morning and the frogs make night music , so that you can see winters at the at the South are a little different from ones we are accustomed to.

The 9th New Jersey have, until a few days been encamped very near us, so that I have made several visits to Chip Hopkinson & Tom Middleton, both beg to be remembered to all at home. Monday, I visited the hospital and saw about seventy five wounded, one of them and the most cheerful of all, is a fellow of the 9th NJ has both legs shot off just below the knees. Dr. Green showed me one bullet which he cut out of a Zouave's back having gone through his body. When I was in there the poor fellow was dying and expired that night. We lost, or that is we had as near as I can ascertain about fifty killed and two hundred wounded, while the rebs had only fifty or sixty. They had so much the advantage that it is wonderful to me, why our men were not repulsed. I have sent on to Marie a small sketch of the battleground, so that you can compare it with those in the papers. Yesterday we escorted the last of the prisoners down to the fort, which our fleet bombarded, and there they embarked on the steamers bound for Elizabeth City. All are released on their taking the oath not to fight until they are regularly exchanged, so that now we have enough to re[a]lease all our men at the South and have some fifteen hundred to spare. Our Lt Col has gone to Richmond relative to the exchange of our men and I have great hopes that Cols Lee & Corcoran will now be released. All of the men spoke very highly of our soldiers, especially our regiment. The Richmond L I Blue, the crack corps of the South, are very pleasant fellows, and say they will never fight again as they were most pleasantly disappointed with our treatment of the prisoners. One of them had a brother at Fort Warren & while there wrote how well he was treated by our regiment, charging him if any of us were taken prisoners to treat us kindly, and now the brother can testify to it. Without doubt the mean released with spread the true reports which their leaders have kindly (?) kept from them, so that other will know we are not such barbarians as is represented. Very often I have seen our fellows sharing their rations with the prisoners when, as was the case for a few days, there was not enough stores on land for all. The fort we visited yesterday bore evidence of our shells, as the trees, some of them nearly two feet in diameter, were cut off as clean as could be done with an axe. Our march yesterday was about eight miles, and that without any dinner, so that we only had two meals all day. "Such is a soldier's life."

Being out of stamps and none for sale here you will have to pay the postage. If a letter is not worth the stamps let me know & I won't write any more, but one thing is certain I will pay the postage on every letter you send as they are worth dollars to me.

We are to remain here for some days yet and when we do move the 24th is to take the advance, I presume our men will have the bayonet charge to make in the next fight and am glad of it as it is the safest way to take a battery, or defeat the rebels. It was shown in the last fight that the Zouaves suffered the least, yet they were in the advance & took the battery at the point of the bayonet.

Direct all letters as I told you. Burnsides Coast Expedition Care Storeship "Marcia Greenfeaf" Old Point Comfort Va. that being the safest way to send them. One of our Color Guard received by the last mail only twenty nine papers, besides several letters. I wish you would send me every paper you can. Gen Burnside says we shall all be at home by next August. I hope it will prove true.

Tell Joe & Mother that this letter will have to do for them as I can't write three all alike. Also tell Joe to give my love to all my friends & say to him that in my drawer he will find an envelope with photographs in it that I wish he would send on to me. Also any others he can scrape up as all of those things will help remind me of home. Kiss the young uns for me & give my love to all at home. I will write as often as possible.

I think it is too bad in that mean scamp Wilson to introduce a bill cutting off our bands thus taking away the only thing the soldiers can enjoy. He won't make many friends by the operation. I presume you will see many big knives in the windows of our stores in the city, which were taken in the fight, if there is any way to send I shall get one as a trophy. However, the next place we go to we shall know how to act & you can bet your life I shall confiscate lively. I might have had the Col's sword but thought I couldn't carry it, but contented myself with his pipe. Speaking of pipes puts me in mind that our boys are crazy digging sweet briar roots to make them of, and if I get any as I o=undoubtedly shall, will make you one.

It is needless to say that I am well as a fellow can't be very sick if he takes care of himself, so now I will bid you goodbye hoping soon to hear from all at home, also to receive many papers, one more farewell.

Your loving brother
Julius

This sheet is secesh paper. One rebel postage stamp from Newbern inside.

Newbern NC Mch 15th/62

Dear John

Bully for us, we have given the rebs an awful thrashing, for particulars see the papers. This is how it came to pass.

Tuesday morning Mch 11th we got upstream & started for Hatteras where we anchored that night. In the evening we recd a mail bringing papers & yours & Mother's letter, for which I am truly thankful. We also recd an account of the doings of the Merrimac Enccon's battery. Was Elisha Turner saved? I hope he came out alright. I tell you what, we determined not to take many prisoners after hearing that she fought under the black flag. Send me a full account of the fight. Wednesday morning very early, we again steamed off and at night were about 15 miles from this city. That morning we were told an advance on the Potomac. Thursday morning we were piled into the boats and taken in tow by the steamer while the gunboats shelled the woods. When within a quarter of a mile of shore we were cast off & then took place one of the most exciting races I ever witnessed, as each regt wanted to plant their colors on the beach first. Our boat in which were our Col, Quarter Master, Adj. & Color guard soon took the lead, the Col helping the men row. Soon we struck the mud & some of the other flags passed us, cheering and being cheered by their men behind. However, they didn't do it that time, for over board we went, above the knees in water. Col S being the first on shore out of the boats, although Gen Burnside I think was the first on the other side of the stream, having run his steamship up to a wharf. Well to make a long story short we march on as the advance guard, one regt. Being thrown out as skirmishers, & the gunboats shelling the woods ahead. It rained all day, so that it took till after dark to march eight miles. The roads were through swamps & in some places the mud was two feet deep and being encumbered with overcoats & blankets our march was very hard. That night all slept as much as possible although it rained hard, still having fires we got along tip top. About 7 yesterday AM our line was formed & we again pushed on in advance as our pickets, nearly half mile ahead, were fired on. Soon we filed out from the road into a piece of open pine woods & while marching along were surprised to hear the report of a cannon & the whizzing of a shell as it passed over our heads. Soon the skirmishers were at it, & as the trees grew smaller we could see a long line of earthworks, full half a mile long & as our flag came in sight a bullet flew by our Color bearer & struck a fellow on my right, a few feet off, in the shoulder. I shall never forget the "thud" as it hit him. In a few minutes we halted on the edge of the grove, & laid down, in which position we remained all through the fight. The shot, shell & grape flew very fast & the heavy firing on our left (we were on the extreme right) told that the others had come up & were doing their best. While we were down, heavy guns on the extreme right of us, swept the edge of the wood & the cleared space in front of the battery, so that a charge was impossible. Col G, the color guard & part of Co I were behind a log into which the bullets struck very fast. It was not long before tremendous gains on our right & in our rear gave us courage as we then knew the steamers were in fact at work. I kept my eyes over the log & saw 9 inch shells bursting in & over the battery. One gun belonging to the 23th Mass on our left did great execution among the rebels, while our own fellows fired as fast as they could load, most of them taking aim at the men at the guns. When the fellows on the right were driven from the battery, we thought it was some of our men, as they ran up into the fight & thus missed some glorious shots. Once a white flag was hoisted on sight of which we all rose up & gave three cheers, when a tremendous volley from some 7 cannon & all their rifles & e was fired. Down we went like lightening & just escaped as they flew over our heads cutting down trees very fast. Our Major was wounded though not badly, the Col slightly with a spent ball, Dan Sargent & Lt Nichols through the shoulder & neck just before this took place. I couldn't make my gun go as the water got into the tube, so laid behind the log listening to the balls as they went over my head. On passed through a small hole under the said log & passed by my side. One fellow back of me was hit as he rose to fire. How long the fight lasted I can't say, as I was nearly asleep part of the time, being very tired. After a while we saw the rebels leaving & on looking

[Note: Letter continues on different stationery.]

to the left of the American flag was seen coming up at the doublequick, Gen Reno having flanked them & made a charge. The enemy didn't wait for them but scampered, while we sent our respects in the shape of rifle balls. We made a rush for the battery, jumped into the ditch & scaled the banks, taking our place at the head of the column. On the bank of the river was a fort mounting 12 large guns, fire of which were pointed so as to sweep our forces, & as shells were bursting just beyond were pushed on so as to let them know we were [illegible]. On mounting the earthworks the sight was splendid, as the steamers were slowing advancing firing shot & shell towards two forts ahead, while the shell flew over our heads after the rebels. We signaled to the gunboats the direction & in a few minutes a shell was fired blowing up one of the forts. It struck fairly in the magazine so that now there was only one fort & the City lay at our mercy. The shot fired at our navy didn't go half way, while our shells burst inside, or on the top. In the fort we were in was the surgeons instruments covered with blood showing that some were hit. We (that is the 24th) captured two silk flags & 12 guns in the fort, while in the battery were six or seven brass rifled guns. After resting a little we again took up our march for the city some 5 miles off. In the battery I saw some 34 or 40 horses belonging to the Artillery killed by our shots, & a few dead men. Most of their killed or wounded were carried off to the cards, & others thrown in the ditch. Blood was seen in any quantities showing that their loss must have been large. In our Company, our 2nd Lieut Horton (the acting Adjutant) was badly wounded, will probably lose his arm if not his life. One fellow had both legs & both hands taken off, while another was hit in the head by a grape shot, taking half his head away. We lost these 2 killed & some 6 or 7 wounded. Co F had 4 killed & about 6 wounded. One fellow in Co K had his head completely taken off by a cannon ball scattering his brains all over the men near him. In our regt we had, I should think some 60 killed & wounded. Although the bullets came very near me yet thank God none hit. Well to proceed. We march on the railroad & up to the opposite side of the city & saw great clouds of black smoke rising from the turpentine factories, while the flames on all sides indicated that the rebels had evacuated & set the city on fire. However not much damage was done, as the negroes put the flames out. The railroad bridge was burnt so that we went on the gunboats & were carried round to the other side of the city. It seems that the enemy retreated to the railroad & were taken on the cars across the river setting the bridges on fire as they passed.

By night we were safe in tents of their camp. By letters & what we could learn their force was 7000 in the camp & large numbers of troops were sent down to the city. We estimated that there were 10.000 in the battery with 7 pieces of artillery & 5 32 & 64 pd guns bearing on us, yet they were whipped by about 4000 as not more than that were engaged, although more were in the rear, but did not take a part. I enclose several letters picked up in here.

I am very tired & can hardly write. Also I am very sleepy. Give my love to all at home. Will write the first chance I get. Don't get worried about me, for although I was wet though for two days & nights, I haven't had any cold as yet & my health is very good.

Yours in haste
Julius

PS Will write you again before long, & give some better account of the action, though you will see enough in the papers

Our Capt remained on the boat as his knee was too lame for him to march, so was not in the fight.

Give my love to all in the families & to all my friends.

Your loving brother
Julius

(Letter fragment)

Camp See Newbern Apl 24th/62

Dear John

Your letter of the 16th reached me late yesterday afternoon and as you have set me the example of answering promptly, I will try to imitate it, although as I have told Jennie & Fan there is absolutely nothing to write, unless, perhaps about that fair damsel whom you have enquired for. In the first place her name is Jennie Miller, second, her father is an officer in the rebel cavalry & lives or did live on a splendid plantation a short distance outside the city. A large vineyard directly opposite the house was destroyed & the posts taken to form part of the fort which Gen Foster has built on the spot where the young was accustomed to gather grapes. Nat Wales sent home her sewing machine, which I could have had, if there had been any way of forwarding it. I have learned a good lesson, so that if we do make a forward movement, shall have a chance to send home some such trophy. Capt Stackpole had her card-case & cards which was presented to him by one of our crowd of foragers. In her room was this picture & a silver pencil case both of which fell to my share of the spoils. Some of the plunderers got small fortunes in valuable jewelry, but I did not go into the houses with such intentions. Round on the floor was a great deal of the girls music & I am sorry I send a lot of it home to Marie, one or two of the band were with us, so sitting down at the piano one of them gave us some of the Dixie music such as "Davis quickstep" & "Dixie war song" &c. The peacock feathers which I sent to the babes came out of the duster. One can form no idea of the plundering of a city unless he sees it done, & it would have astonished a <u>citizen</u> to have witnessed the proceedings of the first few days after our arrival at this place, I say a citizen because a soldier gets so accustomed to ^such awful scenes and deeds that the sacking of a town is regarded as merely a novelty & not to be named in the same day as his other views of blood, or the groans of the wounded & dying. Only think it doesn't seem possible that although I have been a soldier but a short time, that now, no sight would be too horrible for me or would spoil my appetite. On the morning of the fight I saw men eating the rebels breakfasts, which were inside the battery, while within a few feet would be the dead & mangled bodies of those for whom the food was intended while the remnants of the horses belonging to the artillery were scattered over the ground some with a cannon ball through the body, or perhaps one leg shot off, while the poor beasts in many cases trying to get up. Within a very small space, not larger than our own yard, were the bodies of most of the beasts which were killed (some 140 in number) mixed here & there with a few rebels. It was a fearful sight & under any other circumstance would have made my blood run cold. But enough of such bloody talk.

I wish you could hear our band which is now playing "Auld Lang Syne" coming in with "Home, Sweet Home", it is enough to make one thank the day he enlisted to go with the celebrated Gilmon. One thing is certain, the 24th has the beat band which has yet gone to the war,& remember this when our regiment marches through the city. Our drum major will take the crowd down, he is the one called the baby when belonging to Capt Stevensons Company. Yesterday I just got a sight of our flag which one of the men in the regiment has been putting our victories. On the upper stripe, extending from the blue field, is the following in letters of gold. "Roanoke Island", in the white stripe directly underneath is the date February 8th 1862; then comes Newbern & under that March 14th 1862. You cant imagine the effect it has, compared with the dirty & stained old flag, for we don't intend to have it cleaned but [to] remain just as it is, although it could make to look just as good as new. We almost worship that banner & if it should ever be taken by the enemy, rest assured that there won't be many of us left when it is seized, but that good time, for the enemy, will never come.

My boat, as I have written Joe, can be sold for $65. not a cent less. Oil it now & then, and as it is now getting pleasant weather you had better put it outdoors for a day or two, so that it won't mould. You can use my oars and boat if you want to in case of a race, but I thought that Joe might like to try his hand in the "Oi- Theria" in case there should be a race, but (END)

(Letter fragment)

2

don't forget that by the Fourth of July our officers expect to be at home. I must confess that we feel a good deal of anxiety about McClellan before Yorktown; if the Merrimac could only be disposed of the Monitor would do more than 50.000 men could in bringing victory to our side. On McClellans movements depend the ultimate success of the Burnside Expedition, and as we still have room on our flag for <u>three</u> more victories, one or two <u>may</u> be added.

There is a rumor here although none knows the exact truth except out leaders, that Gen Reno has been unsuccessful in his attempts to destroy the locks of the canal from Norfolk. It being ascertained that two small ironclad gunboats were to come down the canal, Gen Burnside determined to prevent any such possibility by destroying the canal, which although blocked up near Elizabeth City, was in such a state that the obstructions could easily be removed, so Gen Reno with several regiment, among them the 21st Mass 9th & 51st NY & 6th NH, marched about 30 miles towards Norfolk, where meeting the enemy he routed them splendidly & drove the secesh behind their intrenchments near the locks. Finding that they were reinforced by 5000 men the destruction of the canal could not be accomplished easily, our troops were withdrawn bringing their [—dead—] wounded, but leaving their dead on the field. About 150 were killed & wounded. Another story is, that the locks were broken & the expedition success ful although 600 lbs of powder had to be thrown away in order to make room for the wounded in the wagons. The 6th NH are said to have acted splendidly in this, their first battle, but the Hawkins Zouaves ran & only a few companies could be brought up. I give this story for what it is worth, as it is just as we heard it, you will see in the papers the true report.

Tell Mother that I want one of my white hankerchiefs sent out, it can be put inside a paper. By the by speaking of papers all those you have sent are received & I am very much obliged for them.

The 18th Mass will not come out here, the error occurred in the number as it was the 17th. Only four regiments have reached us the 17th Mass, 2nd Maryland, 103rd NY & 3rd NY Artillery, one thing is certain we can never advance till more reinforcements arrive, cavalry is needed above all things & I can't see why it isn't sent, when there are so many cavalry regiments doing nothing. If we had had a few hundred of them in the fight here, the rebels would have been slaughtered bu the dozens, or if the artillery had been on hand to pursue the secesh, but both batteries of flying artillery were not landed till the victory was gained.

I saw New Fisher the day we left Roanoke Islands, together with Henry T & Harry VanBrunt, all were looking very well.

Friday morning April 25th/62 8am

Our band is again giving us some of its splendid music, what we should do without it none can tell. How we rejoiced when that bill was laid aside, if it had passed Wilsons name would be considered in the same light of Jeff Davis; ask any of the 500.000 men whether they could dispense with their bands& the answer would be no. How any person could wish it take away from the soldier, his, & almost the only enjoyment they have, is more than I can see, for better stop the mails that take away our bands. If Wilson had been with us at Hatteras & heard the cheers which were given whenever we had the band playing, he would have cursed the day he introduced such a bill into the house, that is if there is a spark of humanity in his body, which I don't believe. One thing is certain the soldiers now regard him as their worst enemy for his efforts to spoil our chief enjoyment.

One of the 9th NJ men told me that Chip Hopkinson went home sick from Roanoke & I haven't seen or heard of him since. Tom Middleton went home wounded.

I would much rather prefer you not to publish any of my letters until I give permission people can always hear enough of the expedition by the papers, as there is where we go for information about what has taken place out here. It does seem strange that you should get the news of what is done by us before we ourselves know any thing about it, unless we participate

if he wants them

in the various events. Give the enclosed cotton seeds to Ned Sumner ^ with my best respects. What a jolly good time you must have had at the Sampsons, it is as you say about enjoying yourself while the war lasts & I am afraid the home guard will use their advantages, so that most of the young ladies will be married before the war is over. How does Bill Chick get along? Are the Fullers still living at the Phoenix? Do you whether the Gardners will come out to D this Summer, or go to some other place.

Give my love to all at home & to my friends in Dedham & elsewhere. I haven't forgotten any of them yet & am looking forward to the good times coming. My health still remains "bang up"

Your loving brother

Julius

[/PS Deliver the enclosed notes according to the address./]

Washington NC June 6th/62

Dear John

I have just a few minutes to write this note as the dispatch boat leaves before long to carry the news of a fight which we had yesterday morning, therefore I shall have to be very brief. It seems that the rebels have appeared before this place lately in force & have killed several of our pickets as our troops were very few in number it was deemed necessary to order the rest of our regt with three pieces of artillery from Newbern and on the night of the 3rd inst we were turned out of our beds, packed knapsacks & started off for the boat. At 4.30 the next morning she sailed for this town & reached our destination that night. At 2.30 yesterday morning the men were ordered to prepare for a march, but owing to the non-arrival of the steamer with provisions the expedition was delayed till 9.15 at which time all of our regt except for Cos C & D numbering about 425, 50 of the cavalry 30 of the artillery with 2 guns left the town to drive the rebels back & if possible to give them a lesson which they would remember. After marching about ten miles the rebels commenced firing from two saw mills and from the tree tops surrounding Lieut Jarves in the ankle so that he will have his leg amputated. The artillery were sent forward & soon opened fire. The enemy could not have chosen a better position as not more than one or two companies could fire at a time at them, while all their force were pouring in vollies at a very short range on us. The place can't be described perfectly although one could give a very good idea by plans or a picture, and as Harper's Artist (Mr [Wise?]) was in the thickest of the fight you will probably see a sketch of it. The two saw mills were a short distance apart connected by a narrow causeway with a swamp on both sides so that it was impossible to flank the rebels. After firing for about three quarters of an hour we charged up to the second mill where the floor had been torn up & after a few planks had been laid the first companies rushed after the retreating for but without being able to overtake them. I passed through the causeway on the other side passing by several dead men, one of them had his head blown from his body. Finding pursuit useless & having accomplished all that could be done, the regt was formed on the road where we had laid during the skirmish & then a short rest being given I walked to the hospital to see the wounded. To my regret the first body was that of George Baxter who was then dead, having passed away to a better world without pain;

of his death

as far as possible the particulars^ I have ascertained from a Corporal of Co H who was beside George when he was wounded and who carried him to the rear. George was hit in the throat, just as he was rising to fire when the action was about half over, & did not speak a word after being struck. Dr Curtis told me that he died without the slightest pain. He looked as if he was only sleeping as there was not the slightest trace of agony on his countenance. I did not have but a few minutes to spare as we were to return immediately to Washington, so cutting off a lock of his hair (which I enclose) for his family, I saw him placed on a cart to be carried back to this town. As the weather is too warm, he cannot be sent home as I should wish, but I will see that the grave is properly marked so that at some future time he can be removed. This is a sad loss to his company as he was one of the best men in it & everyone thought the world of him. Thus it is in battle the best ones are always killed Poor George, I can't realize that he has left us all and gone I trust to another world. I sympathize with his family in their affliction but he could not have died better than he did, fighting bravely till he fell. Enclosed you will find a letter which I took from his pocket and several photographs. As there is none of his friends except myself to look after his effects, I have taken the liberty to examine his knapsack for any small articles of value to his family and found a few things which will be forwarded as soon I can get them to the express office at Newbern. If —- wishes me to do anything else please write at once and I will attend to it if possible. His funeral will probably take place in a day or two and if we are here, I shall attend it.

We arrived at our quarters at 8.45 PM, having marched over twenty miles in less than 11 ½ hours Our loss is seven killed and about a dozen wounded. One of our Co was killed & another wounded. I had several very narrow escapes but "a miss of an inch is as good as mile" so here I am safe but very tired. In a few days I will write more particularly but as my time is short, shall have to say good bye for the present.

Love to all

Your affectionate brother

Julius

PS If I can get a Washington paper with the account of the action it will be mailed to you, so that you can get some better idea of our skirmish, than I can you.

Yours in haste

Julius

Newbern July 8th 1862

Dear Mother

We are here still, notwithstanding the numerous orders to leave and if I am not greatly mistaken we shall remain on our old camping ground for sometime. You have no idea what events have taken place within the last two weeks and to relate all the particulars would take up so much time that I will merely give you a short account of our experiences of "the glorious uncertainties of Gen Burnsides Expedition". In the first place a large quantity of forage has been taken up the railroad and every thing needed for an advance towards Goldsboro was ready. It was generally believed that a long march was before us, but it now turns out that our beloved General has "pulled the wool over the rebels eyes" and will turn up where he is least expected. The 24th was ordered down from Washington in a great hurry and we were to be ready to leave Newbern at 3 o'clk AM the day after our arrival. The hour came and we were not aroused, then the next day the start was to have been made, but that turned out to be a sham. Our next orders were to go to Washington as Foster's division would not take part in the expedition, but you see we are still at old Newbern and are likely to remain here for some weeks although at any moment we may go to Richmond. I wrote a letter to John which I expected would be my last before going into battle, but it now appears by the present indications that many more will be written ere that time arrives. Our Expedition, composing the 2nd & 3rd Divisions in all about eighteen Regiments, sailed from Beaufort & this place on the 2nd inst and the next day the news came that Richmond was taken & Jackson killed, so the steamers were recalled, but not before they had reached Hatteras. Last Friday all the troops were landed but were ordered on board again the next day & sailed Sunday morning, probably for McClellans Army & without doubt Fort Darling will see some of Burnsides Army. Gen Reno requested to have the 24th go, but Foster would not let us as the best troops were left behind to defend the forts here in case of an attack. The 4th I passed very quietly, in camp all the morning and took dinner & tea with Bill Fisher. In the afternoon the chief amusement among the officers at the Hotel was setting the little "nigs" fighting and of all the sport I have yet seen this beats everything. I laughed till my jaws ached fearfully, it was such fun to see the little black beggars butt heads for a dime and surpasses tossing them in blankets. In the evening we had another rousing bonfire and as the rockets shot up through the cloud of black smoke everyone was struck with the magnificent spectacle. Later in the evening we were invited over to the 27ths camp to take part in a species of sham fight showing how Richmond was (or as it should have been not) taken.

with a vengeance

This was a sham ^ and after loafing around till about 11 o'clock I turned in.

Bill Fisher has been very kind to me and I am indebted to him for a great many little articles of luxury, condensed cream, tomatoes, figs, cheese &c which have come from his store. These things go very well out here & are hard to be obtained except at a great price. He is making money very fast and as there isn't much chance to spend it will have a nice little sum before long. If I should be taken sick I know he will see that good care is taken of me, and you can rest assured that I have got a friend now who would do everything in his power to benefit me of his acquaintances.

What boat does Turner sail in? Give him my regards if you should see him. Jennie's letter was received yesterday morning with about a dozen papers. I will write to her in a few days. We heard yesterday morning the McClellan had been badly defeated, but of course did not believe any such news, and last night by the arrival of another boat came the joyful tidings that it turned out a victory. You can form no idea of the various rumors of defeats & victories which are circulated in camp, so that now I don't believe anything until I see the official report.

I am very sorry to hear of Mrs Soules death, John wrote that she was very sick but I did not know that she was dead till you letter came.

Do you have hot weather in Dedham now? Out here the thermometer reaches about 100 degrees in the shade but I don't feel it any more than I did at home, we manage to keep cool someway or another; ice is the thing I miss, but whenever down town I feast on it. Last Saturday we received new tents so that now all can laugh at the rain, before that day whenever it stormed we were drowned out, the boys always said that it rained harder in side than it did outdoors. Has Tip Dannell been out to Dedham yet? If he should come & you hear of it invite him & Ned Davis up to tea for my sake, Tippy was always one of my best friends and many pleasant days have been passed together.

If you cant read this writing its not my fault the table is so rickety that it's impossible to keep it from shaking. I should not be surprised if we took part in some important movement ere many months are over, as Charleston Wilmington have yet to be captured & after Richmond is taken & our other two divisions return, the 1st will be sent with the Monitor & Galena to take the hotbed of secession. I really expect to see that city & only hope we may do our share in the attack.

Give my regards to the Molleys Mrs Rodman & family, & love to all at home. Kiss the "young uns". I am very well and fatter than I ever was before.

Write soon to your
Affectionate son
Julius.

Commonwealth of Massachusetts

Adjutant-General's Office

Boston, Sept. 8th 1862

M[?] Mary Lathrop
Schenectady, New York

Your letter of Sept 5th was received this morning and I have the pleasure to inform you that notice of the promotion of Sergeant Lathrop was telegraphed to the Adjutant General of the United States, and on the 6th of this month Official notice of his promotion was sent to you brother and I have no doubt but that he will be at home in about a fortnight to take his position in the 38th Regiment as first Lieutenant.

Respectfully Yours
MW Schouler
Adjt Genrl

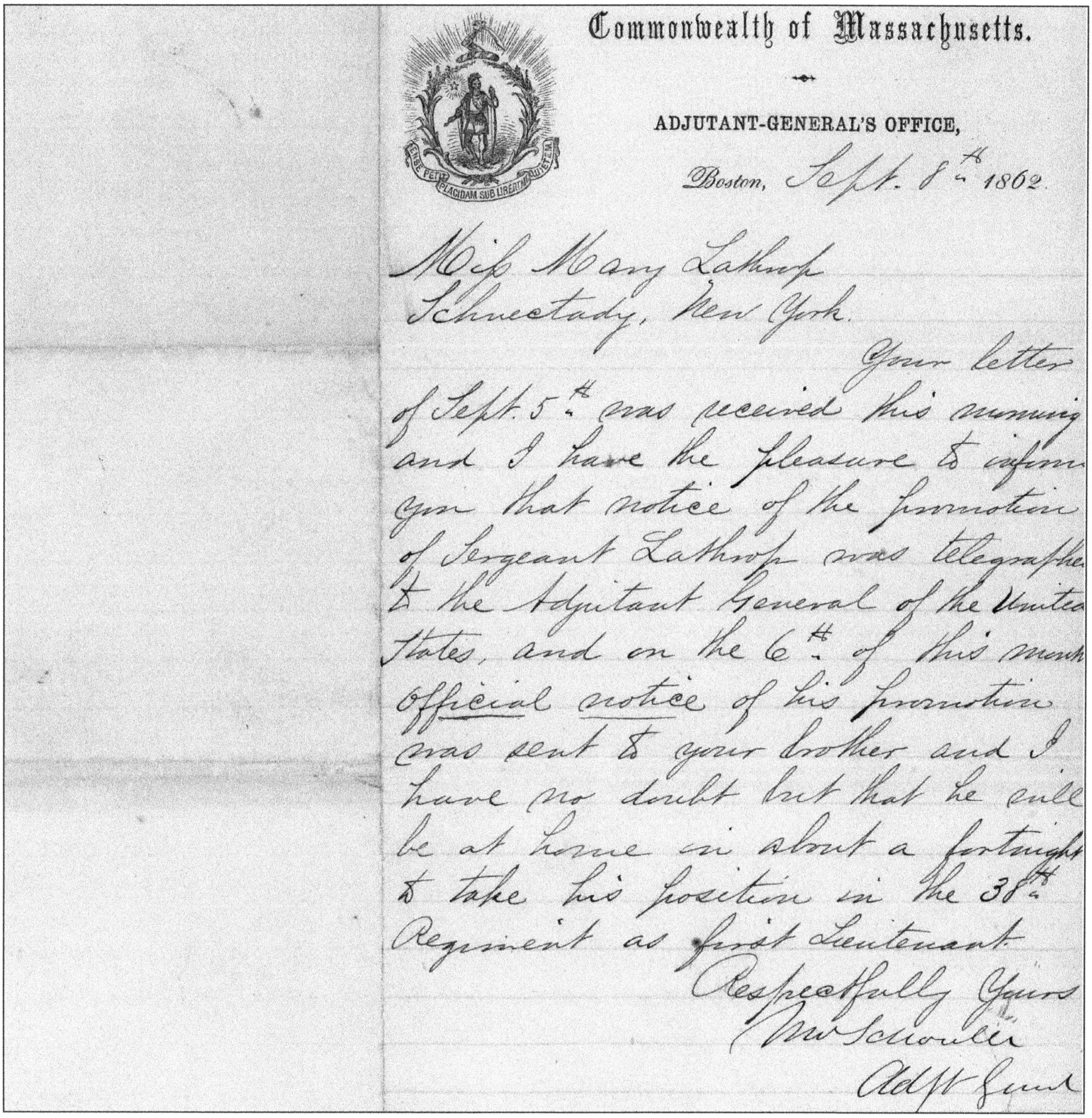

Commonwealth of Massachusetts.

ADJUTANT-GENERAL'S OFFICE,

Boston, Sept. 8th 1862.

Miss Mary Lathrop
Schenectady, New York.

Your letter of Sept. 5th was received this morning and I have the pleasure to inform you that notice of the promotion of Sergeant Lathrop was telegraphed to the Adjutant General of the United States, and on the 6th of this month official notice of his promotion was sent to your brother and I have no doubt but that he will be at home in about a fortnight to take his position in the 38th Regiment as first Lieutenant.

Respectfully Yours
Wm Schouler
Adjt Genl

Julius Lathrop Promotion Sept. 8 1862

Powhattan Md Oct 3rd 1862

My dear Louise

You may think I have forgotten my promise made the last time I saw you , but it is not so, on the contrary after "mature deliberation" so to speak, I turned my company over to the 2nd Lieut this morning till all my letters could be written. The fact is this, since arriving in camp the whole charge of the men has devolved to "this child", Capt Rodman being too unwell to attend to the drilling, and until yesterday morning I have had nobody to assist me, however so far, instead of being "used up" it seems to agree with me, after all there is nothing like Camp life for a young fellow to bring out his energies and make him useful. Don't you think so?

I am having the jolliest time now that one can imagine for a soldier. Up at 5.30 AM, breakfast at 6 ½ Study my lesson in battalion movements and attend to the various little duties around camp till about 10. Then we recite our exercises to the Col and after that is over take the Co on drill till dinner is ready. From 2 o'clk until 5, skirmish or battalion drill employs us and at 5.30 Dress parade closes our daily duties so you can see what precious little time one gets to write home; to be sure there is the evening and even <u>then</u> some one calls, or a crowd are going out visiting or to prayer meetings held about two miles away. The latter place proves the attraction the ladies have as all of us attend pretty regularly, not that we care particularly about the services, but there we meet the pretty girls who live around this section of the Country and then walk home with the "unprotected damsels" which I can assure you is very agreeable to both parties and as the evenings are beautiful just now we seldom return much before midnight. For the last three days I have not been outside of the Camp except while drilling, but to night a large party of the officers are going serenading and have enticed me to join in, which I shall do. It is astonishing that shoulder straps make such a difference with the fair sex, before, I had to stand in the distance & could not speak to a lady, <u>now</u> I can do as I please. You can form no idea what it is to be away from the ladies a long time without seeing but very few as it was at Newbern, and the enjoyment I took when at home is beyond my powers of description. How pleasant it would have been for me if I could have remained there a week longer, but duty comes first and happiness last.

Our camp is delightfully situated on a hill near the Frederic road, about a mile from the little village of Powhattan and seven from Baltimore. Our parade ground slopes gently down to the stream of water in front of the camp and our tents being pitched among the trees makes it very cool during the day. The scenery is beautiful from where I am now sitting in my tent, so different from North Carolina and the trees just putting on their Autumn dress make a glorious contrast to the green hills near by. During the last raid the rebel pickets came within a few miles of the little bridge we are now guarding, and the cannonading of the battle made the ground shake under foot. As the "rebs" are now out of the State, we shall probably break up camp ere many days and leave either for Baltimore or Harper's Ferry so when you write direct to Lieut J M Lathrop Co H 38th Regt Mass. Vols. Baltimore, instead of to this place.

There is no prospect now of our seeing active service before Spring, so I have made up my mind to take things easy, but must confess I am disappointed although a battle isn't a pleasant thing and then ones life is sweet, mine especially to me for never before have my prospects seemed brighter than they do now, and I have determined to get still higher before the war is over. But at the same time, have made up my mind to do my duty or there will be a military funeral in old Dedham and for all I know there may be one yet, still "go it while you are young" is my motto and while life lasts intend to enjoy myself.

Tomorrow or Saturday much to my delight as there is no drill in the afternoon so I can find time to write home, a thing I haven't done for over a week besides that there will be the next day to make calls or receive visitors, for you must know Sunday afternoons we have a crowd of young ladies to see us and witness our dress parade. I must leave off this letter until another time as the bugle will sound for skirmish drill in a few minutes.

Saturday evening.

After reading the Dedham Gazette this evening I determined to finish my letter and will commence with what I did yesterday & last night. Just after leaving off, the regiment was formed for skirmish drill and away we tramped for a field some two miles distant. Arriving I was ordered to take command of two companies & deploy as skirmishers which was done in good style so that I was complimented by the Col for the way it was accomplished. From 2 till after 6 We were out on drill making over <u>seven hours</u> the men were skirmishing yesterday. In the evening a crowd of us started out to go to church, but on arriving there found it locked up, closed for that night, so pushing on soon came to a house where we saw several young ladies in the parlor. For an excuse to call we asked for a drink of water and of course received an invitation to come in which was done, and enjoyed ourselves very much. This morning three of us visited Baltimore to see the "sight" and of course had a great time. I have had some more photographs taken but don't know whether they will be any better than the last lot, or not. I am determined to get one good picture if I have to sit a dozen times. Baltimore is celebrated for its pretty girls and to day being pleasant they turned out in large numbers so that they almost crowded me off the sidewalk; now and then one would "turn up her nose" at the uniform but I usually made up a face at the secesh spirit which will show itself everywhere.

Only think of our fallen resposing under the sod on the bloody field of Sharpsburg and Ned Davis wounded. Little did I think so soon after their departing from home that they would be on the long list of killed and wounded. Remember Louise if I am ever down in the former you must sing in the choir at the funeral, still as I intend to come back safe and sound you may not get the chance.

The Major's now in my tent and has invited me to go out to tea tomorrow evening which I have accepted and as there is a young lady at the house who is very fond of shoulder straps I shall ***[Note: He now begins writing vertically over the previous passages.]*** enjoy myself as much as can be expected on a Sunday evening.

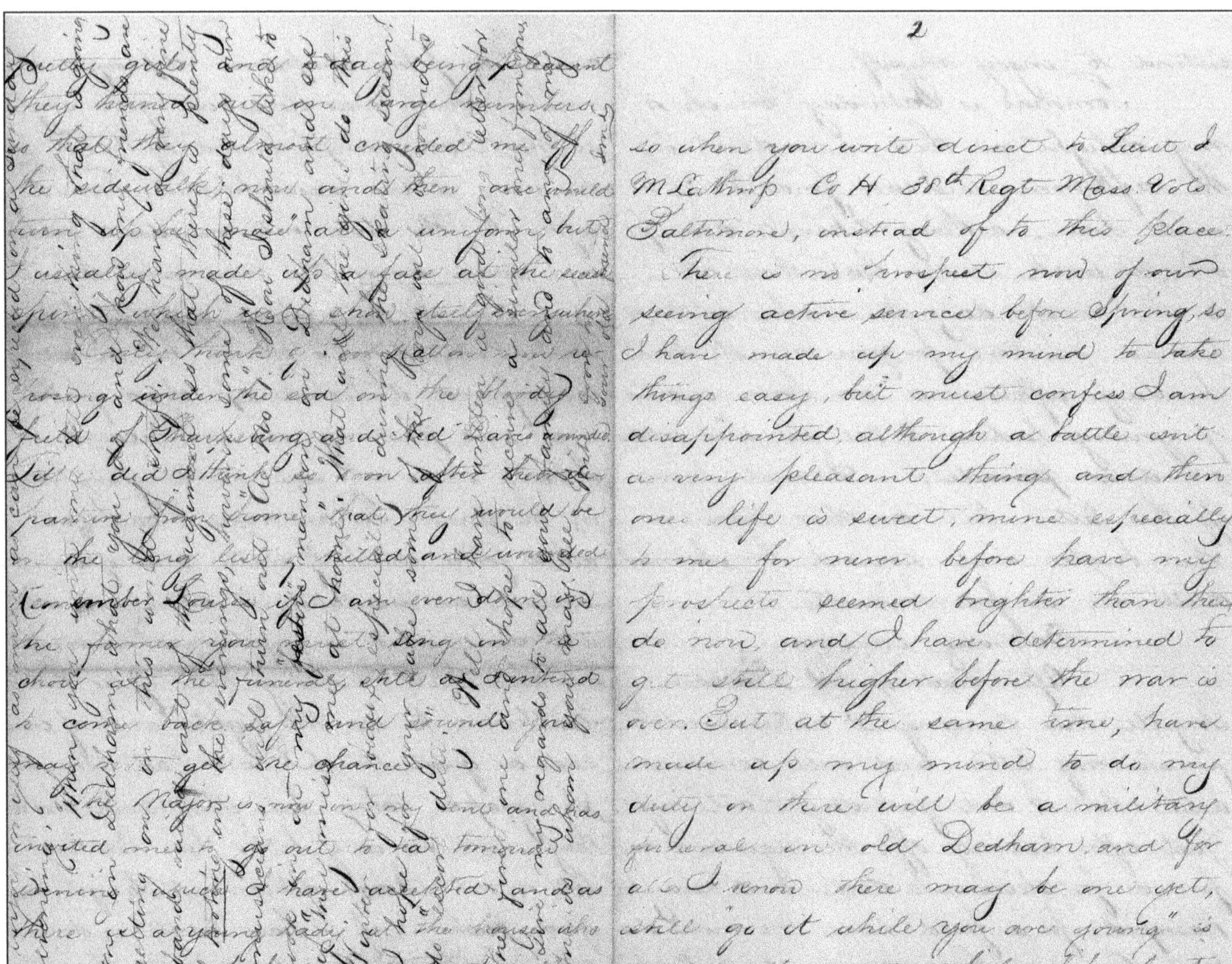
2

so when you write direct to Lieut J M Lathrop Co H 38th Regt Mass Vols Baltimore, instead of to this place.

There is no prospect now of our seeing active service before Spring, so I have made up my mind to take things easy, but must confess I am disappointed although a battle isn't a very pleasant thing and then one's life is sweet, mine especially to me for never before have my prospects seemed brighter than they do now, and I have determined to get still higher before the war is over. But at the same time, have made up my mind to do my duty or there will be a military funeral in old Dedham, and for all I know there may be one yet, still "go it while you are young" is my motto and while life lasts

Julius Lathrop Oct. 3, 1862 cross-hatch.

When you write me tell everything that is going on in Dedham, what you did, and how my friends are getting on in this world of strife. We have a very fine band raised out of the regiment so that there is plenty of rooting in the evening: s, however some of these days our musicians will turn out "A No 1". How I should like to look in at my "festive" mansion in Dedham and see if "they miss me at home". What will the girls do this winter for beaux, especially during the skating season? I hope for your sake that some of the boys will be ordered to "escort duty."

Well I have written a good long letter for the first one and hope to receive a similar one from you. Give my regards to all your family and to any of my friends whom you may see.

Write soon
Your old friend
JML

Camp Emory Nov 5th/62

Dear Mother

We have now received orders to embark tomorrow probably in the morning but still we may not go before the day after. Officers are to carry six days rations so it seems that the Expedition will not go as far as Texas just yet, probably the troops will rendezvous at Fortress Monroe till the Iron Clads are all ready & then we shall see fun. It will be real jolly to witness the monitors banging away, while we stay off at a safe distance till the place is taken, Our destination is not yet known but as the troops are going on board Ocean Steamers, I fancy from what I can learn that we may yet turn up at Hilton Head. Do you know whether Arny Rand is yet there?

Saturday night after writing you I had a touch of the shakes which lasted all night. Tell yesterday I felt rather miserable from the effects of it, but am all right now and have no doubt but that a sea voyage will do me good. Capt Rodman will go home in a few days and he says that just as soon as he gets well enough he intends to visit Dedham & see you. His wife will be with him and I should like to have you take him to Fans & Jennies, also to Mrs. Rodmans as he is her cousin. He has always been very kind to me so do for my sake introduce him to Fan & Gus as I know he will enjoy the visit.

After our pay day, though when it will come I cant see, I intend to send home for a steel vest and a Spencer Repeating rifle, besides sufficient money to replace that which you have drawn from the bank. Don't hesitate to do as I wrote you Saturday because it is a great uncertainty when they will ever get sufficient money to pay us off so draw what there is & make it go as far as possible. I have forgotten just how much I have there, if there is enough to

do so

pay Mr Wakefield $6. For that insurance ^, if there isn't I will send it on as soon as I can raise it, but I shall have to scrape my small change together to buy enough to eat for the next six days, however if worse comes to worst I can live on soldiers rations.

I have packed up everything and written to John & Mollie, if we don't leave before Friday I will send a letter to Joe, Our Col is in command of the 2nd brigade & our regt being chosen to act as Gen Emory's body guard we shant go on board till the very last ones. All the others are safely stowed away on the vessels and we soon shall be off. Don't write till you hear from me. I shall write as soon and as often as possible.

Give my love to all & kisses to the dear little ones. Give my love to any of my friends who may inquire after me, and above all don't worry, as there is no danger where we are going of our being engaged in the attack, the gun boats will do all that.

Your affectionate son
Julius

Nov 6th

PS I see by last nights papers that the 43rd will do in a day or two. I will direct the letter to Newbern. We go on board this afternoon, Stmr Baltic.

Julius

COLEMAN'S EUTAW HOUSE,
Baltimore, ...Nov 10th..........1862

Dear dear Mother

We packed up yesterday noon & after a rather tedious march reached the Union Relief rooms. Here we ascertained that the Steamer is aground and although the baggage is all safe on board, still there is but little prospect of our leaving till the tide is high enough to float her, so we may be here a day or two.

I was paid up to the 1st of Nov from Oct 10th yesterday, & have sent by a Mr W H Jewell of Cambridge, father of one of our lieuts, $40 & requested him to hand the money to Mr Wakefield. He says he knows Mr W so there is no doubt that the money will go all right. Please endorse this amount, when you get it, on the back of that memorandum & I will send the rest after next pay day.

I have ascertained that our destination is Hilton Head and as there is a large body of troops going, there will be some expedition planned against Charleston ere the Winter is out.

What a shame it is that Gen McClellan has been removed. I think it will make trouble in the Army.

Give my love to all &write soon.

Your affectionate Son
Julius

Stmr Baltic Hampton Roads Nov 14th/62

Dear Mother

As there is every prospect of the Expedition remaining round here for some time, I concluded to write to have my direction changed instead of to Washington, send my letters to Gen Emorys Division via Fortress Monroe. I think we shall not leave till the yellow fever at Hilton Head has died away as it isn't safe to go there for the present. When you write let me know whether Arny Rand is at Hilton Head or where he is now as I want to find out. To day the whole regiment has been on shore giving the men a chance to run round and get a breath of fresh air. Tomorrow I shall try to go up to Newport News & see the iron clads but don't know whether I can do it or not. This morning I found out that the gunboat "Chocura" on which Elisha Turner is has gone down the coast, so that I may see him yet. I am sorry as I had made up my mind to make him a visit as I can now go almost everywhere and if the regt stays round here long I shall make a visit to Norfolk. It seems that the old 24th has been in a fight and are in a fair way to see a little more of a same sort still they will come out all right. Write one as soon as you get this & let me know whether you have heard from John. Tell Mollie what my direction is as I write her we were going to Texas & she will be sending letters to that address.

I am enjoying myself very much, as my men are now nicely settled in very comfortable quarters on deck with an awning spread over them in the night time, and I in my pleasant little state room, have a jolly quiet time.

What is going on at home, is it cold there, or are you having warm pleasant weather as we are out here. Yesterday and this afternoon I went all over the Fortress, one can have no idea of what a tremendous place it is till they go inside the walls. I saw the little chapel where Maj Winthrop and Lieut Greble were placed after they were brought back from the Great Bethel fight. Our Col, who was then a Capt in the 3rd Mass Regt, buried Maj Winthrop under military honors. Dear Mother good night. Give my love to all at home and tell Jennie to send me one of little Neddie's photographs by the next mail.

Write soon to
Your affectionate son
Julius

Don't share this letter as I haven't time to write scarcely a line I am in such a hurry & think you will find a great many mistakes. Love to all

Julius

Stmr Baltic Hampton Roads Nov 19th/62

Dear Mother

Why don't some of you write, here I have been sending any quantity of letters & haven't received one in return. Today, I have been to Norfolk and will try to relate my experiences. In the first place four of us officers started at 10.30 in the Stmr City of Hudson from the Wharf of Old Point, & in a few minutes were off Sewalls Point, you have no idea what tremendous batteries the rebs had there and at Craney Island opposite, I wonder they ever evacuated them as none of our war vessels could have passed them without being sunk. Away in the distance, with my little opera glass I could see a dot on the water which was the Monitor, close by was a very large vessel the "New Ironsides" and a short distance from ^ the Galena, these vessels are on the lookout for Merrimac No 2. Off Newport News we saw the masts of the Cumberland sticking out of water and near the beach the remains of the Congress were visable. Just above the battery on Craney Island were several boats at work raising the guns & iron of the Old Merrimac & as the tide was low I could see just a few of her timbers. Coming down from Norfolk we met the Stmr "Commander Barney" one of our old gunboats on the Burnside Expedition, one that I had seen a great many times. Several of the transports that we used last Winter are round here. Norfolk is the dirtiest city I was ever in. I don't wonder they had the yellow fever there so badly a few years ago. The land is very low, but even then if care is taken, there is no necessity of its being unhealthy. I saw some very pretty houses built on the edge of a tremendous mud hole, just the very idea, well, one can see what Southerners are.

We didn't meet any "good Union citizens", all are secesh, even the ladies would swing way out one side so that we shouldn't touch their dresses, and the young fellows tried to make us step to one side to make room on the side walk, but I noticed I had a collision with one young secesh and almost knocked him into the dirty gutter. We walked four abreast so that the females had to crowd by us, it was great fun.

As time was short we couldn't visit the Navy Yard, but saw the ruins in the distance. Coming down I got acquainted with a Mr. T F Wells of Boston, who has the contract of raising all the sunken vessels around here. He succeeded last week in floating the "United States" and is now at work on the Merrimack and Cumberland. He gave us quite an account of what has to be gone through in his undertakings. The Merrimack will be about three months work and 20 tons of gunpowder to blast it. They expect to get $150,000 worth of iron and copper etc. The Cumberland is to be raised whole and it will be several months before she floats. In the safe of the Congress (another sunken Union ship) is $40,000 in gold, to be raised as soon as possible. In all, Mr. Wells estimates there is a million and a half dollars worth of materials which he expects to raise and of this he gets 45 per cent, so that he will make a big fortune out of it.

He has promised to leave with Mr. Wakefield a piece of the Merrimack for me, but I don't think he will trouble himself about it, still as he is going home this week, he may keep his promise.

Our fleet may sail in a few days & it may wait a week or two, like the General doesn't know himself. I have now an idea that we are to go with Banks, but cant tell for a certainty. The regt goes on shore nearly every day. Monday we march to Hampton and of all the deserted places this beats everything I have seen. Perhaps you may remember it was burnt by the rebs over a year ago & now nothing is left but the ruins of the houses. It is a perfect picture of war in all its desolation. In the pretty little church yard are some tomb stones bearing the date of 1701. Just outside the town are the earthworks built by the 3rd & 4th Mass Regts a year ago last Spring to defend the place, but when those regts went home Hampton was left unguarded, & one night a small party of sesech came in & set all the buildings on fire. What have you heard from John lately? I wish someone would write & let me know what is going on
nowadays
as I don't get a letter nor even a paper from home ^ & you have no idea how discouraged I am getting. It don't pay to send letters & not get an answer. I have written several times about some money I sent home & would like to hear whether it ever reached its destination.

I am in good health & am getting quite fat out here. The weather is splendid, quite warm & comfortable.

Yours in haste
Julius

PS Tell Mrs Turner that the Chocura has sailed for Hatteras so I shan't see Elisha. Abel Smith was in the 3rd NY Regt & now stationed at Fortress Monroe, but he has gone home to accept a Colonelcy in a new regt.

Stmr Baltic Hampton Roads Nov 26th, 62

Dear Mother

I received your last letter about a week ago the day after writing one to you. Since that time we have been lying idle as before, though I have improved the time by going to Norfolk Navy Yard and Newport News. I don't feel much like writing this afternoon as I am rather lazy and sleepy the result of reading all the morning through. I must tell you about the Monitor and the New Ironsides (the actual name of the vessel) both of which I passed quite near yesterday. For a long time I have intended visiting Newport News so as to get a good view of the ironclads lying just off the point and not till yesterday did I succeed in getting to the boat in person. The Ironsides is a tremendous vessel worth several Monitors as she carries sixteen eleven inch guns besides two hundred-pound Parrotts, where the little Monitor has two eleven inch cannon. The latter vessel lies quite close to the wharf so that all of us had a fine chance to "take observations". The immense dents on the turret and sides show where the shots from the Merrimack and Fort Darling hit her. One of these, the largest, was from the Fort and can be seen nearly a half mile off. It shows how much the little vessel can stand, and how powerful the rebel guns are. I was rather astonished to find her such an insignificant little thing as my ideas were on a larger scale although very little shows out of the water, there is a great deal of room below. After coming back and while waiting on the wharf at Old Point, who should come across but Elisha Turner, he has grown so fat that I hardly recognized him. I found out a day or two ago that his vessel had got back from her cruise and was up the James River, so I wrote him a note to call & see me. At Beaufort he saw all the Dedham boys in the 43rd & says all are very well & in fine spirits. He represents Joe as looking very tough & hearty. We had a good long talk about home matters & old times & left him at dark with the promise of calling on me as soon as possible. Tell Mrs Turner that I have never seen him looking so well in my life, he has gained about twenty five pounds since leaving home.

I wish to pay for all that insurance so pocket the $6. & say no more about it. Tomorrow we are to have a big Thanksgiving dinner on board ship & shall enjoy it I have no doubt, as sea air gives us an appetite. The weather is quite warm so that overcoats are not needed except nights. My health is very good indeed and I am gaining every day. I never saw such a scene of desolation in all my life as there is at the Norfolk Navy Yard, over fifty buildings, each of them about twice as
at home
long as the depot ^ and six or eight times the size, are in ruins. It was a sad sight to look at those tremendous piles of ruins. I brought away a few pieces of burnt slate as relics to

remember my visit. There is some talk of our destination being changed, but I can't tell & have now given up guessing. The impression is now, Mobile or New Orleans, and that there is no doubt but that we shall go with Banks. We are liable to sail any day, however I will write as soon as we know.

Love to all
Your Affectionate Son
Julius

On board Stmr "Baltic" Hampton Roads Nov 28th, 62

Dear John

I have expecting to hear from you every day but have a t last concluded to write you my whereabouts and give my right address. For the last two weeks and a half we have been lying on this steamer, the regt going on shore every pleasant day for drill and target practice, and coming on board at night again. You can easily see what a nuisance this is, it takes up all a fellows time & then with school of instruction in the evening it leaves us but few minutes to write in. I recd a letter from Mother yesterday in which she says you are acting Major. Bully for you old boy. I hope yet to see the gold leaf on your shoulder. Since our arrival at this place I have made two visits to Norfolk, one to the Navy Yard and one to Newport News, and all for nothing as the boats round here carry shoulder straps for free. Quite good idea don't you think so? I have had a good view of all the ironclads off Newport News including the little Monitor, and the tremendous "New Ironsides", just think what an armament this vessel has, sixteen eleven inch guns and two two hundred pound Parrots, 1600 pounds of iron at a single broadside.

By the way Georgie Dennie wishes me to ask you for that photograph which you promised her and says when she receives yours she will send one of hers. Did you ever get those wristers I sent them by mail one at a time three or four weeks ago. How do you like Burnside? Isn't it strange that I should fight under him in one state and now you in another. I have the greatest faith in him & always said sooner or later he would be at the head of the Army.

I wish you could have been here at dinner yesterday, Capt Cumstock the commander of this splendid boat invited us to dine with him, and I can tell you what we had a gay time. All sorts of dishes that you could think of, with pies, puddings & pastry besides nuts, raisons & apples. Parker couldn't have got up a much better dinner, and as champagne was thrown in we drank many toasts. My 2nd Lieut & myself have a tip top stateroom, good beds with <u>white</u> sheets, marble wash stand and everything in the best style. You have been on one of the Ocean Steamers so you can tell just about how we are situated. Meals are .50¢ each but I only get dinner, breakfast I make up from crackers & butter, smoked beef & sardines. Two meals a day only.

I got a letter from Joe to night in which he says they have gone into camp near the Trent. All are well. He wasn't sea sick at all during the voyage. At Beaufort they all saw Elisha Turner who was down there on the "Chocura", and Wednesday I met him on the wharf & had quite a long talk. He weights about 25 lbs more than he ever did at home. I wish I was in the Navy, a fellow can have a decent assortment of clothes on board ship, without living in fear of being obliged to send all home. Joe is a commissary Sergt for the Co so that he doesn't have to stand guard, I am glad of this as it will be a good thing for him. The 24th are on picket duty nine miles from the City.

How does Hill get along with the Company, I know he wasn't liked by any of the Dedham boys in the 18th. How I wish I was in your Regt, although this is a very good one still I should much rather be where there are men from old Dedham. When you get to be Major you can give me the command of your Company, I shall be well instructed in everything requisite for a Co commander by that time, as I have had almost the entire charge of mine ever since coming to the regt. Even when Capt R was on duty he never went to Co drill, only on battalion, & dress parade. Major R says he intends that I shall have a command of my own some of these days although I only rank seventh of the 1st Lieuts. How was it about Nat Wales & the sesech damsels, is there any truth in that report? I know he had a weakness that way when I used to know him.

When you write me send that photograph which was promised to me some time ago. You have never acknowledged the rect. of the second one I sent. Did you ever get it? Write me always as often as possible directing Gen Emorys Division via Fortress Monroe. Without the slightest doubt we are going under Banks way down South either to Charleston or Mobile, the latter place I think, though there is a prospect of turning up at New Orleans, still its hard to guess. There are now in this vicinity some 15.000 men lying idle and at Suffolk there are about 25.000 more. Why can't we all be put on the road to Richmond via Petersburg, I believe we could easily take it. There's some prospect of our whole expedition making a dash in that direction and it would not surprise me to receive orders any day. Here government is paying $1500. a day for this boat & nothing is being done, besides the Baltic are nearly a dozen large vessels filled with troops close by, among them the "Spaulding". Several of the gunboats which were with the Burnside Expedition are to go on this one, Whether the Passaic forms one of our fleet I am unable to state, though Elisha Turner says she is to take the Galena's place while the latter boat goes home for repairs.

Frank Loring sends his best regards to you, he has just recovered from jaundice and is once more on duty with the regiment. What sort of fellow is your 2nd Lieut. I heard some bad reports about his actions at South Mountain, somewhat the way Geo Perry did at Balls Bluff. You seem to be unlucky in your officers. I hope Johnny Cobb will soon get a commission as he always was a tip top fellow. Give him my best regards, & say to him that turner saw his brother Sam & states he is looking quite tough& hearty. Commodore Van Brunt is going on our Expedition, thia is with Banks and I have no doubt that he will have command of the vessel forces. I wish the Ironsides could be one of the vessels as she is a beauty, although I have more faith in the Monitor standing shot better as she has been tried. I saw part of the old Merrimac sticking out of water just above Craney Island & off Newport News the burnt timbers of the Congress & the masts of the Cumberland. The are raising the latter vessel.

Well old boy good bye. I must close as its time for our school. Write soon.

Yours in great haste
Julius

(Letter fragment)

3

Tactics, we shan't make that famous raid as the Major is ordered back to his regt which is to be changed in to Cavalry and sent into Texas. It will be a great disappointment to us if we are to leave here for Baton Rouge just as we have everything fixed comfortable for the Winter.

Last week news came that a large force of rebs were coming to clean us out, but as we had built two pretty strong forts, one each side of the town, didn't care for the enemy. It now turns out that some 6000 or 8000 men were really marching on this place and were to have made the attack early Sunday morning, but at last came to the conclusion it wouldn't play although they outnumbered us ten to one. Let them come, it will rare fun for us I can assure you.

Don't you wish you were out here now? Yesterday, and in fact for nearly a week, I have sat here reading and writing with my doors and windows wide open although about ten days ago ice formed nearly an inch thick. The grass and trees are quite green and the garden almost as bright as ever. The musquitoes still trouble me at night so that I sleep with my bar down, a few more frosts and then they will sing no more around my bed till next Spring.

I presume you are having skating and possibly sleighing at home now, what a difference. I prefer this kind of weather to the North just now, though I should enjoy a skate once more. Oranges are plenty, won't you come out here and eat some with me? I know you will enjoy yourself, you can ride out almost any day wearing summer clothes instead of furs which you now have to put on whenever you move outside the house.

I go out duck shooting nearly everyday and always get enough to supply my table, there's more sport here gunning than at home, plenty of game and one does not have to tramp but a few rods to get it.

Thanksgiving the officers of our detachment got up a big dinner and having invited several of the western officers to partake with us, sat down to the festive board at 4 P.M. I have no idea how many kinds of meat &c we had, but there was everything one could wish for just a good a dinner as Parker could have got up for such an occasion. We had all kinds of wines &c egg-nog in abundance and the way we all drank toasts was a caution to the total abstinence society. I never enjoyed anything better in my life than I did that dinner and when we all rose from the table at none o'clock, determined to have another such time Christmas and New Year. I only hope you passed your Thanksgiving as happily as I did mine.

I see by the papers that you have been singing at some of Gillmore's concerts when you see Mr. G again give him my regards. I presume though he has forgotten my by this time.

And now good bye. Remember me to any of my friends whom you may see. Write soon to.

Yours truly
Julius

PS If you want to see how I now look get Joe to show you one of the pictures which I have recently sent home.

Stmr Baltic Dec 3rd 1862

My dear Mother

I have written several letter to you lately and am now commencing another in answer to one I received today. Yesterday at the very last minute I managed to buy a valise just the thing I wanted and so concluded to send my trunk home. You know that a recent order prohibits trunks in the Army and now in a hurry I can catch up my two valises and run at a minutes notice. As I shan't want my sash I put it in the trunk with other things you had better take them all out and get a list of just what is inside and then put everything back as I shall need some of them ere six months are out. Being well supplied with a good pair of shoes and boots I let that pair go home. Next Spring I will write for a box and let you know what I shall need and you can send them by Express. Don't buy thatvalise but you purchase a good steel lined vest and if any presents are given Christmas I should like a knife, a good strong one. Capt Rodman promised to go to Dedham and see you & will probably be there ere many days. He will be staying at Mr Frank Welds in Jamaica Plains. I think he will take a bundle to me with greatest pleasure. Yesterday afternoon after sending my trunk home, two of us jumped on a tug expecting she was going to the Baltic, but judge our surprise when the Capt told us after after leaving the wharf that he was to pull off a streamer that [cross out] was aground and it might take all night. It was rather chilly and then having no overcoats on we felt decidedly cool. On board the officers were eating dinner while we two hungry mortals stood shivering in the little tug. The Col told us to hurry back as the fleet might sail at any moment, however there we were in a fix. After an hours pulling the steamer slowly came off much to our great joy and at last we were carried on board the "Baltic".

There is no secrets now about our going with Banks, it is a settled fact and in future direct to me Gen Emerys Division, Banks Expedition via <u>New York,</u> and tell of the others to address me the same way. [Ilegible name or title] Backus did not reach Baltimore till after we had started away from there. I was very much disappoint in not seeing him but it was my usual luck.

Do you remember last year when our Expedition started from here I wrote a letter to you and sent it back by Pilot when off the Capes? I shall do so with this as my last chance of sending it on shore by the Postmaster is gone. Our destination I think will be Pensacola and from there to Mobile. Butler will have a hand in the pie also, the mortar fleet are collecting these and the gunboats at New Orleans, while great operations have commence down the River. Our Lt Col Wardwell resigned to day on account of ill health but I think ere many weeks are out he will be on some Generals staff. He was a very good fighter and would have handled us in good style in battle. Major Rodman will now be Lt Col. He has risen very fast commencing as 2nd Lieut, but being a good friend of the Governors has got up. I have written you two letter this month, this makes the third besides eight last month so I don't think you ought to complain, still they have all been very short. In future I shall have to write very long ones but it won't be often. I can send them, you will get one if anybody does. It is expected we shall start tomorrow, in fact I've heard the General say so himself. About fifteen steamers carrying ten thousand men will join Banks outside and will go on our way to carry desolation into the Southern homes. They brought upon themselves so let them bear it as best they can. When you write John and Joe as well as Mary tell them my address as letters will come sooner that way than any other. I sent the key of my trunk and the Express receipt in my letter yesterday to you and hope they will get along all right. I have with me now three sets of underclothing, plenty of stockings handkerchiefs and everything I require for an age. You remember that little valise I carried away, it is now stuffed full and another the size of Johns well packed. My blankets are strapped outside so that they don't take up any room. I paid the freight through to Boston so it won't be more than a quarter more to pay on it. Send the bundle which is inside by Express to Plymouth but don't pay the freight on it. If you should ever see Mary Gardner put her in mind that she promised me a pair of wristers which I have never received yet. I am afraid she has forgotten her promise. I sent those Emmie Gardner made for John to him by mail and he received them all right.

Dec 4th 3 PM

At last we are off. At 10 this morning the first vessels started and soon all but the Baltic and our convoy the Augusta were on their way. Ours being the flag ship we had to wait till the last moment and at the time I am writing we are passing the fortress. Now don't worry because you will not receive a letter for an age, I will write by the first boat North. We expect to meet Banks' fleet outside the Capes as most of them sailed two days ago from New York. I expected to receive a letter from Jennie today but didn't, tell her to send Freddies photograph by the first mail. By the way I wish you would say to Mr Wakefield that I should like to get that rifle as soon as it can be made, I want very fine sights put upon it and at least 500 cartridges. Several of the officers will get them after seeing mine, & if he speaks to Mr Clark about it I think he would hurry it up. Don't forget this as I am in a hurry for it. Give my best love to all at home and to all my friends with many kisses to the babies.

Your affectionate son

Julius

Ship Island Dec 24th 1862

Dear Mother

I should have written to you a few days ago had we not been expecting to receive letters from home very soon, but as there is no prospect of getting the mail till we reach New Orleans, I will scribble off a few words & write again as soon as possible. Contrary to our expectation the regt has been compelled to stay on this miserable place and I cant tell when we shall get away, not for some days at least, so we shall have to pass our Christmas in rather a stupid style. This afternoon is very warm so that I feel really uncomfortable just now, don't you wish you had a little of it? Nothing by sand in every direction, one cant even eat with any pleasure as the prevailing element is sure to get mixed up with ones food. When I spoke in my last letter of there being no living things on the Island I forgot the spiders and the land crabs and at the upper part of the place there are a few alligators in the lagoon. Some of the soldiers killed one of these beasts two or three days ago.

Don't expect me to give you much of a letter this time as there is nothing at all to write about, all we do is to drill & sleep, pass away the time as easy as possible. When I get to some place where there are a few more troops and [houses?] I may be able to fill out several pages, but while on this Island I don't feel like attempting to write. We had some glorious oranges a few days ago from Havana, but at Army prices .5¢a piece, on board the boat they were sold for .60 a hundred but we couldn't get off the vessel so had to buy of speculators.

In the storm off Hatteras one of the New York regts has an officer washed overboard & drowned, he was on one of the smaller steamers. It is said that Banks rather surprised Butler as nobody in New Orleans has the slightest idea we were coming down to this part of the country. I suppose that Butler has gone home ere this.

What is going on at home? Write all the news & send me a paper as often as you can, we don't get many done here I can assure you. I must close now dear Mother as the mail is going right off, however I will write a longer letter next time.

Give my best love to all & many kisses to the babies.

Yours in great haste
Julius

Stmr "Northern Light" off Carrollton
La Jan 1st 1863

My dearest Mother

You may think I have forgotten all at home by not receiving more letters, but so many things prevented sending mails before this one, that it could not be helped. I wrote we were at Shi Island and were expecting to leave every day for New Orleans, but to our disgust the regt was compelled to remain there fifteen days in that horrid sand. How I did hate it. No mails from home all this time and no means of getting any news to amount to anything, plenty of rumors only. Monday night last the regt embarked on this boat and passed up the river yesterday arriving off NO a little after dark. If I had time to day I would give you a long account of what we saw along the route, but being in an awful hurry to get this thing in the first mail tomorrow I shall have to put off my story till we are fairly settled in Camp.

the 31st Dec

Last evening ^ after just four weeks absence from Fortress Monroe, the long expected mail was received and for my ration I got a jolly long letter from John dated Dec 3rd, one from Jennie same date, and a fourth from Molly Dec 7th. It was the best New Year's present I ever received and just at midnight too. As it is impossible to answer them all at once I wish you would tell Fran and Gus that I will give them good long answers ere many days. I want to run down to the city first to see if I cant find our sesech cousins and will give the results of my investigations in my next. Where we are now is only five miles from the city and a railroad & horse cars every few minutes so that its glorious. As far as I can judge Carrollton is a very pretty place, but perhaps was may not like after all, anyway it is better than Ship Island. Here we shall without doubt stay a long time with the rest of our brigade. Any quantity of oranges 10 for a quarter, and the weather is so warm that while I am writing there are girls sitting on the bank without bonnets and all have their parasols. What do you think of that?

By the way it appears that Burnside was badly beaten and I learn this morning that poor Sidney Willard was killed. What a sad thing for his many friends, how John will miss him. He and I were situated very much alike, he had Willard to do everything for him, and I had Lt Col Rodman who treats me splendidly, somehow or other he has always taken great interest in me ever since I joined the regt and I can appreciate how John feels as I well know it would be a bad thing for me if Lt Col R was killed.

We have lived in great style on this boat, far better than on the Baltic which was tip top there. How I wish I could have time to give you a long description of my wanderings since leaving the Baltic, but if I wrote everything to you I should be hard up for news for the others.

By the way, as I can't write to John my full account of the trip out, having forgotten a great many of the incidents of the voyage, I want you to send on to him the letter I sent you as it will be of some interest to him. What a pity it is that the 35th could not have gone with us, we live in such great style, but some of these days I expect to come down to the fine thing.

We are to be paid off in a few days and I shall send home $125. By the way please tell Mr W that he need not do anything more about that rifle till I write for it, as I may be able to get a good one out here. Capt R is expected here almost every day. Did you see him? There is a tremendous fleet of war vessels at New Orleans now waiting for iron clads before attacking Vicksburg which will come off in the course of a month or so. I doubt if we are sent up there as there is every prospect of staying at C. a longtime.

Tell Jennie to send on those photographs in the next mail as I do want to see the little ones <u>so</u> <u>much,</u> don't forget.

What cold weather you must be having at home now and here I am "in clover", grass green in all the gardens. You ought to see the orange plantations, I never saw such a fine sight of its kind as I did coming up the River, it was by all means the best trip I ever took in my life and you can rest assured I enjoyed it exceedingly. Send me a list of the killed & wounded in any fight which may take place as we don't get many papers but those published in N.O. I haven't any idea scarcely of the fight at Fredericksburg, all I have seen is a few lines which showed me we were badly beaten losing 10.000.

We dear Mother I must close, promising to write again in a few days. Give my best love to all & wish them a Merry Christmas & a happy New Year. I am getting along tip top now.

Yours in great haste

Julius

(Letter fragment)

1

Camp Kearney Carrollton La Jan 14th/63

Dear Mother

As there is a mail going tomorrow I concluded this evening to write a few lines, so that you can keep the run of what takes place in this part of the Sunny South, but before commencing I must say that there is precious little to fill up a letter. In the first place we are under marching orders and expect in a few days to do to Baton Rouge where the ground is higher and consequently much healthier that where the regt is at present encamped. The slightest rain floods our street so that it is decidedly unpleasant, and now that the river is rising I presume there will be a great deal of water in the many ditches which surround us. Camp Kearney is a pretty place but we can't stand the mud, so away we shall go ere many days; another bad feature of this place is the drinking water, you never saw such dirty stuff, still there is nothing else, but at Baton Rouge it is said that water is splendid. Well on the whole I am rather glad of it, anything so that the men will be in a better condition than at present, I lost one of my company night before last & buried him yesterday back of our parade ground. He had been sick almost all of the time since leaving Mass so that I have seen but little of him. My health is tip top and if you could only see how I eat you would be surprised, the little sickness at Ship Island will do me a great deal of good and already I feel the benefits of it. Oh dear! If the wind didn't blow so hard I could write you <u>quite</u> a long letter to night, but the candle flickers so much that I can hardly see; a tent isn't quite as close as our house is in Dedham, the wind finds plenty of place to whistle through and the consequence is my light is none of the best, however I won't give up quite yet. This morning I enclosed a paper to you showing what kind of news we get out here, of course coming through sesech sources it is all one sided & sometimes makes us feel quite blue. For instance there was a report last Sunday at headquarters that Rosecrans had surrendered his whole army, but late Northern papers say he whipped Bragg. Foster's raid on Goldsboro we heard of some time ago but the rebs said they defeated him at Kinston, I didn't believe this & now the truth comes out. We are all very sorry to hear of the loss of the Monitor as we can't afford such a thing, well I am in high hopes to get the news of the attack on some Southern port by the iron clads ere many days have passed, but mails travel slowly to us. If the Alabama doesn't capture the steamer which left here last Saturday, I shall feel quite gay that my greenbacks escape the rebel clutches. By the way I have bought a splendid carbine out here so that you can tell Mr Wakefield I shan't want one of those rifles, the one I now have belonged to an officer who was hard up so he sold it cheap, now if my vest only comes I shall be better equipped to go into a fight, carrying a rifle instead of a sword so the rebs won't have a chance to distinguish me from my men.

Capt Rodman is expected to arrive in one of the boats now overdue, but I hardly think he will join us before we leave this Camp. Just think of it, today has been so warm and in fact every day is the same, that we drill in shirt sleeves and the horrid musquitoes come buzzing around at night. Don't you wish you had a little of the heat? I can't realize that there is ice and snow at home; when I am writing especially, as then it seems as if I was talking to you and instead of being 2000 miles away, I imagine myself once more in old Dedham. Well I should like to skate once or twice, but after all I prefer this weather to the ice & snow especially when you can live on oranges. By the way speaking of living puts me in mind that I must tell you how "this child" survives & what he eats, so that you can see if I don't go on the right principle. Breakfast usually consists of toast & butter, coffee, roast potatoes & sometimes Baked fish, meat we can't get. Dinner & supper are about the same except that rice & milk form part of my second meal and with oranges several times a day I continue to grow fat, nearly all my rations come from the Commissary so that it is rather cheap, not over 30 or 40 cents per day for grub. I wish John and Joe could always share my rations, my little "nig" being a very good cook I get along nicely only having the trouble of telling him what I want for each day.

The Fowlers I haven't been able to trace having visited the city but once, if you could find out from the Bordentown people what part of the city they live in, I might hunt them up, still I don't think they would care about seeing a Union soldier in their house. (END)

Camp Kearney Jan 16th/63

Dear Mother

You see by the heading that we are still in the same camp although having received marching orders several days ago, I thought our departure had come, for between 1 & 2 o'clock word was sent to the officers to prepare for an instant march. I got up, roused the men, saw that each one had one hundred rounds of cartridges and two days rations, blankets rolled & knapsacks packed. As the order was to go on light marching order all baggage except blankets & overcoats was to be left behind in the tents, so after packing up my clothes and seeing that my rifle cartridges were ready, I sat down to a breakfast of toast & coffee at 2.30 AM. You know when I wrote night before last that it was very hot but yesterday & to day a "norther" has been blowing and the consequence is extremely cold weather so that ice formed last night, however my nig had a jolly hot fire in the stove and by this means I kept comfortable. Everything was soon in readiness and there we waited for further orders till at last about 7 o'clock we were told to box up the additional 60 rounds of cartridges & return them. This noon it turned out somewhat as I supposed and here we still remain though liable to receive instructions to report at Baton Rouge. It seems that the rebs attacked Gen Weitzel at Berwicks Bay about forty miles distant, and last night word was received by Gen Banks that they had whipped the rebs in fine style, but more ammunition was needed and he might require some reinforcements, that is the reason we were called up at such an early hour. As yet but little information can be obtained about the victory except that Lt T McKean Buchanon, a son I think of the paymaster Elisha Turner was with, is killed in the gunboat fight & was buried to day. The rebel gunboat Cotton is captured.

Today a steamer arrived from the North bringing papers of the 6th so that we have good news from Vicksburg though I think it queer that we haven't heard more about it through Southern sources. The "290" a day or two ago sunk the gunboat Hatteras off Galveston and I am afraid she will capture the steamer which carried my greenbacks home, if that is the case I shall be $125 out. Now if you don't receive the money soon, that is in the course of a week or two, draw out of the bank enough to pay yourself, as it is very uncertain whether that money ever reaches its destination in safety. It seems that Rosencrans gave Bragg a glorious thrashing at Murfreesboro, quite a different story from what we heard from the rebel papers, its hard to believe anything down here. I have expected Capt Rodman by every boat but somehow or other he hasn't yet made his appearance. I hope he will come before we go into active service as I shall want my steel vest, with that and my rifle I will be very well equipped.

To night is much warmer than it was this afternoon as the wind has gone down, that's the way here one day hot & the next perhaps very cold. Tomorrow morning we are to get our mail and I presume I shall hear from some of you but am afraid the mail North will close ere we get the other letters, however I can answer them next time.

Saturday morning

Last night being still was very cold, quite a heavy frost appeared on the ground this morning & ice could be seen till after 10 o'clock. You would laugh to see the people round here bundled up, at home we would call it a clear frosty morning, but inhabitants say it is uncommonly cold weather & so they go round with all the clothes they can pile on.

That Buchanan who was killed day before yesterday. I now think is a nephew of the Charlestown family & not a son.

Now dear Mother I must bid you goodbye as the mail soon closes and I want to send this letter by todays boat.

Give my love to all & don't forget the young ones. I am in good health & spirits.

You affectionate son
Julius

Camp Kearney Feb 20th/63

Dear Mother

I have got a few minutes to spare before the mail closes so if a short letter will be acceptable I will write one. We arrived in Camp this morning from our Expedition without having gone beyond Plaquemine, as the reconnaissance down the Bayou showed it blocked up with drift wood thus preventing any boats passing so nothing was left for the brigade but to make a masterly advance on Carrollton and here I am safe & sound without having seen one of thee enemy. As it is very important tht the supplies of salt &c should be cut off from Port Hudson, I think another Expedition will soon retun to Plaquemine & cut their way through the Bayou. It seems that the rebs have got a tremendous salt mine in Texas from which the Confederacy is supplied & if we can get by this timber raft, we can capture or destroy their only line of communication. It will be a big thing if successful & now that every particular is known the next time it will be turn out better for us. A mail has arrived & was sent up the river yesterday we passing it on the way, I presume I shall be hear from home when it gets to us, at least I hope so. My vest is a little large but will do very well, won't you send me an account of everything you have paid out for me in your next letter as I want to know how my affairs stand at home; by the time this reaches you you can give me a list of what you paid out for the thins I sent for several weeks since. Enclosed you will find a five cent piece of sesech money which I promised to Jennie but forgot to send in her letter.

You have no idea of the suffering there is around the South, I never had the slightest idea of it till I went on this Expedition, whole families at the little village of Plaquemine would starve if it were not for our troops and the husbands & brothers are fighting against us. At the North things are different, but here the poor people have to take it, husbands & sons dragged out of their houses & forced to join the Army then if they try to escape are shot. Provisions can't be bought except at tremendous prices, flour <u>$100.</u> <u>a</u> <u>barrel,</u> sugar .50¢ a pound & such like charges. Ain't you glad you don't live here? Pay day will soon be round & I shall send home some more money, out of it you can take enough to pay yourself for what you have advanced for me. Mortar boats have gone up & soon the attack on Port Hudson will commence. We shant be in it. Love to all.

Write soon
In haste
Julius

Camp Kearney March 4th/63

My dear Mother

Today we received orders to be ready to leave at a moments notice, and as the whole division, of some fifteen regiments, is going I presume Baton Rouge will be our destination, still we may go to some other place. New gunboats carrying tremendous guns have arrived and gone up the River, something like a dozen mortar boats are already there, and when the attack on Port Hudson does take place the rebs will find more than they expect. Either a fight will come off at Baton Rouge by the sesech attacking our forces, else we shall commence operations at Port Hudson within the next fortnight. It is about time this Expedition saw some service and the men are eager for it. There is one consolation the land at Baton Rouge is a great deal higher than where the regiment is now encamped and much healthier. To day I was
of the Co
transferred to Co I as the Capt ^ has sent in his resignation & it has been accepted so I was ordered to take command of the Co. Within a very short time if nothing happens I shall receive my commission as Capt there, so in future you can address me Lieut JML &c Co "I" 38th &c instead of Co "H". Tell Fan and Jennie as well as John & Joe of the change of address. I was in Co I 24th, John is in Co I 35th and now I am in Co I 38th. This is a change which will bring me my two bars very soon. Capt Rodman is quite well & sends his regards.

By the way in the bundle which I sent home were the following articles. 3 Undershirts. 1 Pr drawers. 1 Pr gloves. 1 Pkge letters. 1 Pkge photographs. 1 Needle Case. 1 Rubber pillow. The latter Joe bought for me at Walls Co Milk and Congress Sts. The first time I used it I found it leaked but could not discover the place where the air came out. I wish you would get Mr W. to take it in to the store & state the circumstances, perhaps they can mend it, but as it is now I cant get any use of it. Some of these days I shall send home for more things & if it is fixed I can use it.

I haven't been to see Mary yet as we have so much to attend to that I can't find time to run down town. Some of these days I will give her your message. I received the Dedham Gazette from Fan, and one from Dr Burgess by the last mail. I wish you would thank the Doctor for me as it was very kind in him to send papers as he does by every mail, he doesn't forget that there is a Dedham boy way out here. I have seen Paine & Bobby Everett lately, they were both in my class at scholl.

I am very well & in tip top spirits. Give my love to all & write soon to

You affectionate son
Julius

March 5th 2 PM I received a letter from Fan to day dated Feb 3rd. Where it has been all this time I don't know, however I will answer it as soon as I have time. Good bye
Love to all

In haste
Julius

Camp 5 miles from Baton Rouge Mch 19/63

My dear Mother

In my last I stated that we were about to start for Port Hudson and presumed there would be some [tall?] fighting ere writing again. Friday Mch 13th the fleet started up the river and in the afternoon and evening Gen Grovers division took the lead on the land forces followed by Emorys div. with Augurs in reserve, the total number artillery & cavalry included was about 20.000. That night our division camped near the spot where we now are and early the next morning pushed on. At noon Saturday the troops formed line of battle in a large field within six or seven miles of Port Hudson, while Grover went on to reconnoiter. Everyone expected an attack as it was reported the rebs had come out on another road for that purpose, but after remaining twenty four hours the order was given to fall back as the enemy would not be enticed into the position we prepared for them. Nims battery threw shell into Port Hudson and the advance cavalry had a little skirmish resulting in the death of the Major of the 2nd RI Cavalry & Col Clark of Gen Banks losing a leg. Saturday afternoon the mortar fleet commenced shelling the batteries and now other shells could be seen bursting over Port Hudson, but the woods prevented us from viewing the grand spectacle. All the men had little shelter tents, so we managed to get along quite nicely at night. About 12 o'clock I was aroused by heavy broadsides being fired on the river and looking out of my tent discovered that there were lively times on the river. The noise was tremendous as the fleet and batteries were firing! We could tell by the flashes just where the rebs were and our own boats. About 1 AM a light was seen flashing down which we supposed was a fireraft but it turned out afterwards to be the Steam frigate Mississippi which got aground under the batteries & finding she could not be saved was set on fire. After a while the boat floated down stream, her guns going off as they became heated, and by 5 AM she blew up with a fearful report making the ground shake for miles. After this all was great. At noon an order came from Gen Banks stating that the object of the Expedition was accomplished the Hartford & the Albatross having passed the batteries and gone up the river to Vicksburg. So we made a masterly advance to our old camp 5 miles from Baton Rouge. The men don't think the object was accomplished as I believe the intention was to take Port Hudson but Banks discovered his old enemy Stonewall Jackson was there with a tremendous body of troops so he fell back. That night it rained in torrents and you can see how dry we must have been sleeping on the wet ground. Tuesday noon Gen Dwight's brigade with Ingrahams & Goodings pushed out some distance, the latter (our brigade) being held in reserve & secured an immense amount of cotton, some say nearly a million dollars worth, I don't believe it is as much as this, but it was a nice little pile. During the night the rebs came down & drove in the two brigades to us but did not show a fight. Yesterday morning we marched back to this place, and here we are safe & sound without seeing a fight. I presume the troops will return to Baton Rouge in a few days, as the rebs don't care to attack us nor we them. Two or three nights ago I received a letter from you dated the 19th Feb & one from Jennie the 22nd. I will answer hers as soon as we get settled somewhere.

You have no idea how disappointed all of us have been in not taking Port Hudson, but it was soon evident that our force was not half large enough to do anything. However we vented our spite of the cattle, poultry, &c of the sesech farmers and great was the number of both did we kill for food.

The weather is very hot and as marching isn't pleasant I shall be glad when the troops return to Baton Rouge, I must now close as the postmaster leaves soon and will write more particularly next time. Love to all

You affectionate son
Julius

Tell Jennie that the shoulder straps must be bullion not double but single row and to be Captains. I will send money as soon as we are paid off to pay for them. Government owes me about $400. yet I cant get trusted for a cent for provisions.

Camp near Baton Rouge Mch 24th,63

Dear Mother

I received your of the 5th a few minutes ago & will now answer it. Direct my box Lt JL Lathrop Co I 38th Regt Mass Vols Gen Emorys Division Banks Expedition New Orleans La. and it will come all straight no matter where the regt may be. Pay the freight & take duplicate receipts one of which you can keep & send the other to me. If the regt is off it will be forwarded to us or kept in New Orleans till they know where we are. You say that in your last you give an account of the money spent, I haven't received this letter & wish you would mail me another account when you write again. Send on anything that will keep, hard gingerbread, some pineapple chees, cocoa & milk as all of these things are very high out here. For instance at a little eating house you have to pay .50¢ for a plate or two of buck wheat cakes and that not half enough for a meal. I never saw anything like the prices four or five times higher than at home. A few water crackers would eat very well out here. No matter about any medicine as the doctors have a supply & I can get anything I want from them. You might get a bottle of cherry brandy, take off the label and put a new one on calling it some kind of medicine. This is the only way it can come through as all kinds of liquors are stopped in NY, but by this plan you can get it along. We shall be paid off as soon as some money can be sent out here & I will then forward a supply to settle all my bills. I wrote to Jennie a few days ago about my box and if you have not started it, put in what I have mentioned in this letter to you.

The whole division is order to be ready to move I presume to Berwicks Bay in Texas to cut off communication from Port Hudson, but it is very doubtful if we leave, however don't delay the box on this account. There won't be any fighting here unless the rebs attack as as both sides are afraid of each other & don't dare to move outside of their fortifications. I shall write a letter to Fan telling her what is going on round here so that this one will be short. In a few days or as soon as I can answer a few letters I will write to you again.

In haste
You Affectionate Son
Julius

[Note: Written on the back panel in the center of a tri-fold is "Mrs. M M Lathrop Present"]

Camp near Baton Rouge Mch 31st/63

My dear Mother

It is getting pretty well along towards midnight, still I must write these few lines as we are ordered to move in the morning, where I can't tell but presume to Donaldsonville and from there to Berwicks Bay in Texas. Gen Grovers division started last week for that place and tomorrow the first of our division leaves. Possibly we may go to Carrollton but I hardly think that is likely. However I will just as soon as possible after getting to our destination, which may be one or two weeks, everything depends on the amount of marching.

Send on my box anyway, I wrote all about it last week. I hope some of you have a knife for me as I lost the one Fan gave me just before enlisting in the 24th a few days ago when on the last tramp. There is nothing new around here except that the Richmond and several of the mortar boats have returned from Port Hudson and will be of service if the rebs attack the place as I think they will just as soon as we are gone. We haven't been paid off yet, but expect to very soon as the [needful?] has arrived at last much to our great joy. Today we have had a brigade review and inspection which has taken up much of my time, and to night the orders came to start. Having to issue clothing and look after a great many other things, I haven't had a minute to spare to devote to you till after everything was done, so of course can't write a long letter it being very late, and I feel decidedly tired. For the past two days, the weather has been remarkably cold, overcoats being needed even on drills, but Saturday was very hot indeed. This sudden change takes right hold of a fellow I can assure you. Have you heard anything of Ned Davis yet, whether he is still in the hospital or at home! I should like to know so that I can write to the poor fellow. I haven't heard a word from John for an age, why doesn't he write? I presume he must have some time now that he is in Camp at Newport News.

Did you go to any of Gilmore's Concerts? If I had known of it in season I would have treated you to the music. How I should like to hear that band once more, it used to sound so good at Newbern. I see Bobby Everett now & then, he was one of my old schoolmates & is now in the 3rd Mass a clerk for Actg Brg Gen Gooding.

Well dear Mother I must now close. Give lots of love to all & kiss the little ones for me.

I am very well & in fine spirits. Shall write soon.

In haste
Julius

Algiers Apl 8th 1863

Dear Mother

Last night I received a letter from you date the 3rd March and one from Jennie of the 8th. Both of these letters by mistake were put into the Newbern mail as they were stamped on the outside "Not at Newbern NC" this accounts for the delay. I wrote you a few days ago in which I stated that I had sent home $100. & enclosed the rect & a $20. Bill. I now send $10. More to be put to my credit as we are going to day to Brashear City near Berwicks Bay and I shan't need the money there. Last night I got a long letter from Joe, he writes that he has seen Mrs Handy. I see by the papers that the 35th has gone out West to Burnside, Wouldn't it be queer if we were to meet some of these days. I wouldn't be much surprised if we went up the river next Summer, but at present there is some work to be done down here as the rebs came near drowning out Weitzel by cutting the levee & letting in the water. We shall probably be held in reserve while Weitzel & Grover attack the enemy. Plenty of men on our side over 20.000 while the rebs are said to have only 15.000. We have a great many batteries of field guns besides 30 pound parrots and as far as I can learn have the advantage over the enemy. Our baggage is to be carried but our tents are to be left behind barring only the little shelters to live in, this looks as if there was a probability of returning to this place ere long, still one can't tell much about it, everything depends on the result of the attack. One brigade is now marching along to the cars and to night or tomorrow morning we shall be on our way. The railroad runs directly to Brashear City so there wont be much marching I hope.

I am in tip top spirits and health and ready for any emergency.

Have you seen the article about the 38th in the Traveller of Mch 24th or 25th? Gen Banks order is in it.

I am in such a hurry that I can't write more now. Will answer Jennie's letter as soon as we get into Camp again. With much love to all & kisses to the babies.

I remain
Your affect. Son
Julius

Berwick City Apl 10th/63

Dear Mother

Our division with Grovers & Weitzels are now at this place having occupied it yesterday morning without a shot being fired, and to day rumor says we are to start for the Red River a distance of 180 or 200 miles. If we should tramp this distance (which I doubt) you wont hear from me for some time, still I hardly believe that the expedition will be gone many days. Yesterday AM we left Algiers & reached Brashear City just opposite. The rebs fled when Weitzel crossed and I suppose the troops will follow them up. Just think of the clover being six or eight inches high on the field where we are now encamped and along the railroad in some places people are making hay. Blackberries and Strawberries are plenty. The former selling at 10¢ a quart. I saw lots of alligators in the bayous yesterday the first live ones I have ever seen.

Don't you wish you had some blackberries? They are very good indeed. I am in hopes to come across some on our march. How I wish I had my shoes as these I now as these I now have on are almost worn out, and am afraid they won't last the tramp if we are gone a great while. Well write as soon as possible but I don't want you to worry if you shouldn't hear for some days.

Love to all

In haste
Julius

(Letter fragment)

In camp about 70 miles from Berwick City
April 18th 1863

Dear Mother

This is the first chance I have had to write a line since leaving Berwick and now don't know when it can go, still I will scribble a little just to show you how I am situated. One week ago started on the march and on Sunday came up with the enemy on Pattersonville or rather I should say beyond. For about two hours the artillery were at work just to try the strength of the rebs so as to know what we could do with them on the morrow. You have heard ere this of the loss of our gunboat Diana taken in Bayou Teche so I wont write about that affair & only mentioned this to show what part this boat had to do with the plans. Late in the afternoon she came down with the rebel flag flying & opened on our batteries with slight injury to some of the men & horses. At dark firing ceased & during the night our troops were busy building a pontoon bridge across the bayou & planting the siege guns, 20 pound parrotts so as to open a heavy fire in the morning on the gunboats. At daybreak our brigade crossed the pontoon under a fire of shell but without injury, & then advanced into an open canefield sending out the 31st Mass as skirmishers who soon found the enemy. All the morning the batteries were playing on the Diana with fearful effect & when she retired from the fight, turned their guns on the rebel earthworks. At 1 PM the infantry advanced on the other side & the 38th were sent to relive the 31st. My Co was on the left of the line of skirmishers with two Companies in reserve & about 2 PM although under fire we were ordered to advance into the enemy's works. Here was perfect madness, the idea of a regiment of skirmishers charging a long line of batteries & rifle pits, but it was all our Brig Gen knew. It seems that until the rebs opened on my Co with artillery that he did not know their force was so strong. I sent back word that they were in earthworks & could see plainly with my glass the guns in position & wanted our battery to commence firing so as to draw their attention while we rushed forward. It did no good however so on went the regt the rebs running from the ditches where they were to the rifle pits. A perfect storm of shell & bullets greeted us still the men pushed on as cool as if on drill & the line they kept was perfect. On the opposite side they cheered to see us go ahead so beautifully & not until a shell had burst in the reserve killing a Capt & a private besides wounding eight more that we began to think of stopping. The shot tore through my line fearfully but being deployed I only had one or two wounded & one killed. Getting my Co in line I ordered them to lie down & commence firing at the gunners. Here the regt remained for three hours till ammunition was gone & being relieved by the 51st Mass we retired to fill up cartridge boxes. Two men were hit at my side & once a shell covered me with dirt so that I couldn't see for a few minutes. It was lively times I tell you.

In my Co I had two killed and four wounded. In the regt were some 31 wounded & six killed. At dark we rested shot & shell coming in now & then. On our right was the 156th NY & against us as I learned the next day were the celebrated Crescent regt from New Orleans (the one [-ins] Fowler used to belong to), the 33rd & 18th La besides two regts of Texas rangers. It is no use for me to write or to try to write the particulars of the action as I know nothing about it except what transpired around me, so you must not take what I say to be the true report of the battle. The papers will have the news. Don't show the letter to any one as I haven't had time to think much what I have written, but will send a letter report some other time when I can collect my scattered thoughts. Next morning at daylight found the rebs had got enough of it & cleared out during the night so pushed on taking some few prisoners, all of whom complimented us very highly on the way we advanced & the accuracy of the firing. I fired my rifle about forty times with good aim but don't know whether I hit anyone or not. We were so near that the rebs could see my sword and knew that that I was an officer, our boys could hear them yelling to each other "theres an officer kill him &c. The bullets came pretty near I tell you. Tuesday afternoon the rebs blew up the Diana & between some of our gunboats the Queen of the West was destroyed with several other gunboats. At dark marched through Franklin, one of the prettiest towns I have ever seen. Plenty of people were out in the piazza (END)

Opelousas La April 26th 1863

My dear Fan

I received your letters of the 15th & 29th a few days ago and will try to answer them. You complain that I do not write to you by looking back to me memorandum I find that I have sent letters to you as follows Jan 4 Mch 5 Mch 27th. It is my intention to always keep up my correspondence when in the field writing is a nuisance having no table or books to use as a desk then shelter tents are so small that one cannot sit upright in them, thus you see I am not to be blamed if I do not give all my spare time to my friends.

Since arriving at this place the Army has had a rest except now and then sending out parties for cotton and sugar, a vast amount of which has been collected, estimated at $5.000.000. this may be rather large but before we get through it will exceed that sum. It is a pity that they do not give prize money to the soldiers else we might have a nice little share of it, but others than the soldiers will get the benefit, someone will make a fortune I think. Here the troops will remain till all the cotton, sugar and molasses have been confiscated and sent to New Orleans, then another move and another rest. This time the Red River may be our destination, but we may not go further than Washington, some seven miles distant as there is easy communication to Baton Rouge and New Orleans from that place. I only hope that our regt may go into summer quarters ere many weeks have passed as the weather is now very hot. I saw quite a number of men sunstroke in our late tramp and the longer we wait the more dangerous it will be to travel in the heat of the day.

The island on which our brigade had the fight is called "Cow Island", rather a different place from another spot of the same name near good old Dedham. I shall never visit the pond and island again without thinking of the battle of Fort Bisland where the 38th received its first baptism of blood, and where I lost a good friend in poor Capt Gault. We do not miss him yet but shall as soon as a camp is formed. His tent being next to mine I was in there as much as I my own and many a pleasant hour have I passed in his society. Poor fellow he now sleeps his last sleep having fallen in the early part of the fight while leading his men to the attack.

Last night the 8th regulars left us having received orders to depart for the North so a salute of 34 guns was fired. These are the brave fellows who were surrendered by the rascal Twiggs two years ago and now having taken some revenge on the Texans are to go home on furlough. You have read of their sufferings I presume and how nobly the remained true to the old flag. I don't think they took a great many prisoners, I shouldn't if I was in their place.

Capt Wade was dishonorably dismissed the service as he refused to appear before a military board of examination. The story about the Company giving him a gold watch on departing is untrue they gave him a silver one before leaving Lynnfield. My Co is I, Slade had K, but Col Rodman though it best to give me the former as things had gone very slack under its former commander and I could receive more credit for bringing it up to the proper standard than by taking a well disciplined one. The men fought splendidly in their first action, but it was necessary to expose myself more this time than I shall again, as they were more willing to advance if they saw me near them than if I had kept in my place (in the rear) when skirmishing. They now know there is not so much danger as one would think, besides seeing that I am willing to share everything with them, while I am well acquainted with my Co and can tell just what can be expected.

To be sure that battle was not a severe one, but it proved that the 38th could uphold the honor of the old Massachusetts and from our Brig, and Gen Emory we have received compliments. Gen Banks says in his address "You have marched during the last twenty days nearly 300 miles, defeated the best troops of the South in four engagements, taken over 2000 prisoners and ten pieces of artillery, besides destroying gunboats, foundries and salt works." I cant see the 300 miles but different parts of the Army have defeated the enemy four five times. The gunboats reduced the fortifications at Bute la Rose taking all the survivors prisoners. Truly we have accomplished great results and have struck a blow at the rebellion which will make it shake from Richmond to the Rio Grande.

I am expecting news from John and am rather anxious to hear some word from him. Joe wrote a full account of his chase after the rebs which was very interesting I assure you. I never want you to publish any of my letters as whenever I write it is always in a hurry and I never stop to think what I say, especially when in such a situation as this. I have just heard that our division is to march to Washington and trust this true as we shall be better off there than in this old cornfield where the sun beats down in all its fury.

Hos is Johnnie? I trust he is well be this time. No invitation was sent to me to attend reunion. I think they must have forgotten that a certain High Schooler is away out here in Dixie, one who always remembers the former reunions with the greatest pleasure and hopes once more after the war is over to again meet old school mates on that occasion. I presume but very few of the old scholars, especially the boys were present, another set has grown up one which I am not acquainted with, and by the time I get home again, those who were in the lower schools when the war first broke out will be happy members of our great "institution". Dear me how old I am getting to be 23 the fifth of next month, it doesn't seem possible yet it must be so.

Perhaps you might like to know how we have lived during this Expedition and as I have not said much about it in my letters to Jennie, I will do so now. In the first place we will take it after a long days march when arriving in camp tired out and hungry to say nothing of the quantity of soil which we carry in our faces and clothes for the dust lies several inches deep in the woods and so many men, guns and the long baggage train rather increases the quantity and quality, At times one cant see ten feet ahead so thick are the clouds, therefore at night the first thing we think of is water, then wood for fires; as everyman carries a tin dipper, such as I brought home with me , you can see in each fire quite a collection of them well colored by long use in the smoke. In these dippers the men make their coffee and supper once over (that is if they are fortunate enough to have anything to eat as at times on this march not half rations have been issued) if the weather is pleasant no tents are pitched, but wrapping up in a rubber blanket we soon sleep as soundly as if no enemy was near or that we were at home. How sweet is sleep especially after a hard days march. At 4.30 the next morning reveille is sounded and by 6 the column starts halting now and then for a few minutes rest. Our dinner is eaten anyway and consisting of hard tack no time is allowed unless a burnt bridge should prevent the troops from crossing till repaired only once were we fortunate enough to have sufficient time to make coffee in at noon. By 6 o'clock if there are fields near the halt is made and in less than five minutes hundreds of fires can be seen in almost every direction. Lucky are those who can have the night to rest in, as picket and guard take quite a number from every regiment. Once twice have I had my rest spoiled by picket, the first night out and the last one we marched. On arriving at this place before having a chance to get a mouthful to eat or drink, I was sent out in charge of the pickets of the whole brigade and as it took several hours to place them you can imagine I was slightly tired by the time I reached the reserve. But little sleep visited me that night, the only rest I had was by lying down in the furrows of the field, ploughed ground remember, with no blankets & nothing but my rubber coat to cover me. Strange to say the next day when relieved I felt in tip top spirits not the least tired though when I took off my shoes the stocking came too so caked were they with mud that they had stuck fast to the leather. I nearly forgot to tell you that both shoes gave out on the first days tramp and the consequence was that I have been nearly barefoot ever since till a day or two ago when I had them patched up a little. These shoes I had worn ever since Joe sent them to me in Newbern last Summer and I have rather felt the want of those in the box which I sent home in my trunk last December. I am in hopes to get the box ere many weeks have passed or as soon as we get into summer quarters. By destroying their forts and gunboats the rivers and bayous are observed and we now have, or will have in a few days easy communication with New Orleans. Our food since leaving Algiers has been varied from nothing to one hard bread a meal to honey, geese, ducks, chickens &c but now all that I live on is coffee and cornbread or hardtack now and then foraging a piece of meat or a fowl, but the country is so cleared out that not a thing can be obtained. People here are starving the last barrel of flour sold a short time ago for $180. As usual as soon a we enter a place the inhabitants refuse to take any money but greenbacks and as the men have collected quite a pile of sesech money they feel indignant that they cant pass it for eggs, chickens &c so refuse to part with their hard earned $73 a month and the consequence is get what they can without paying a "red". I think that when we three boys get home we can keep you all busy telling stories about our campaigns and adventures in the South, Sunday evenings for instance when we are taking tea with you. Shouldn't I like to be there now, or this evening sitting down to a good supper instead of the humble meal of corncake and coffee. Thanks to a good appetite I relish this which at home I would not touch [illegible] how hungry I was. Not once on long weary march did I drop out of my place and never yet since enlisting have I given up, the thought of that has often prevented me from dropping down by the roadside used up, luckily my health is tip top else I never should have withstood the tramp. Long exposure to the sun has given me a color nearly as dark as a mulatto and I do not think you would recognize me now if I should return suddenly. What is going on in Dedham? I presume Annie Summers future will be married as soon as possible as the married men are not to be conscripted until the bachelors are all gone. What a lot of marriage there will be among those who compose the homeguard. If we ever get back to New Orleans I will have some pictures taken & send you one, but the prospects now are not very encouraging. I hardly think we shall see the crescent City for some months to come.

I should like to sit down and give you an account of our march of the place we have passed through and the hardship endured, but it is in vain that I try to collect my scattered thoughts, nothing but rest and a wall tent will bring me back to a sense of the trials and incidents of the campaign. You can read the papers, these will be more interesting than anything I can write, and not till the true reports come back from the North shall I be able to tell what has been accomplished. Enclosed is a slip of paper which I cut from one the rebs published in this city, it gives the destruction of the Queen of the West. We will see the paper was published on common wallpaper it shows that the rebs are hard up for that material and at some future day will be a relic of secessiondom.

Nobody was injured in my old Co but a color corporal had his head blown off. Co H had the colors and this fellow belonged to another Co his brains were spattered over everyone near him. Just think of the water we had to drink at times, hundreds of men washing in it & yet this was all we could get & glad it was no worse. One becomes accustomed to these things very quickly and don't mind dirt in the least. We are satisfied with what we get.

Now dear Fan good bye. If I don't write very often it is not my fault I assure you. Give my love to all & many kisses to the babies.

Your affectionate brother
Julius

Papers come along all right. Send the Dedham Gazette as often as possible.

Opelousas La May 2nd 1863

My dear Mother

I received a letter from you a few days ago dated April 2nd and at the same time got two from Jennie and one from Fan. Enclosed is a letter from Jennie which was intended for John, but being directed to this regt instead of the 35th I opened it to see if it might not have been written to me. You will all have to be very particular when addressing either of us as just one figure will make a great difference as to the destination.

All are waiting with the greatest anxiety about Charleston, as the news which came was rather doubtful; on the way up an official notice was sent to Gen Banks that the place was taken, also that the rebels had attacked Baton Rouge and had been defeated but it now seems that neither of these reports are true; how disappointed the soldiers get sometimes, news comes which all believe but the next mail brings the different account and oftentimes the reverse instead of the victory. Gen Foster it seems is in a tight place, but from what I know of the country and the situation in which he is placed I cannot but help believing that he will turn it out all right in the end. Things remain quiet out here except that Gen Dwight had a slight skirmish some miles from here yesterday, but the rebels are so displeased that it will take many a week to get the conscripts together once more, and even then they won't feel much like attacking the main army, as they well know the large number of guns we have on our side.

Yesterday afternoon while looking over some old papers I found the enclosed receipt which I supposed had been sent to you a long time ago. Has the money ever got along, together with $30. In two different letters? If so write or get Fan to do it for you, also to send on what the boots cost and the shoulder straps with freight &c. Tell Fan and Jennie that I will answer their letters in a few days, but as everything is so dull now that it would be of no use to write to more than one of you at a time. I am very sorry to hear that you have been so sick and hoe that your illness is all gone by this time. How is John? Give him my best love and say to him for me that I am very much grieved to hear that he has been so unwell just at a time too when his regiment was about to enter upon a new campaign and in another part of the country. I trust that he will soon be well enough to rejoin the 35th. When Joe returns as he will within the next two months, if spared, <u>keep him at home,</u> don't let him go away again as I think that John and I are enough from one family and besides he is needed at home to look after our things, there is my gun it needs cleaning, and my wherry ought to be oiled and exposed to the air else it will get covered with mould. I shall never give my consent to his going out again, no matter how good a chance he may have or how high a position he may attain. He has had his share of fighting and ought to be satisfied but I know he wont.

The weather here is awfully hot, and living in little shelter tents we don't get much comfort I assure you, how long we are to remain here thus is more than I can tell but do not think many weeks, as the heat is growing too intense to march during the day. I do hope that the 1st of June will find us in a summer camp and in a place where I can get at my box, which is very much needed my shirts being in rags and my shoes full of holes, so that marching is very uncomfortable in dust or mud. Last night I was on picket so of course feel decidedly tired and sleepy this morning. By the way, I received a letter from Belzy a day or two ago, how does the youth look? It seems that he has joined the temperance society, which from his accounts rather takes a pile of money to belong to, more than I should care about giving, though if he is sick he gets the benefit of it, but if he is not sick more than I have been during the past year and a half, but precious little recompense would he receive. About $6.000.000 worth of cotton and sugar has been sent to New Orleans and yet still it comes long lines of wagons filled with it are to be seen every night coming along inside the lines. I hope it will soon be all gathered so that we can move from this place as our camp is not very pleasantly situated, and all are tired of the dreary spot. By the way can't that knife be sent by mail? I will pay all of the postage as I want it very much having lost mine when on the Port Hudson expedition and am now without any. My shoulder straps I am expecting ere many days and trust they will be what I want. I should have bought them in New Orleans but things are so awfully dear that it was better to wait till I could send home for them. Since writing to you last I have ascertained that the name of the spot where we had our fight is "Carr Island" and the fort is called "Bisland". I should like to visit Carr Island near Dedham much better than the Carr Island out here.

Currier is ready so I must close. Give my love to all, with kisses to the babies and hoping soon to hear of yours and John's recovery.

I remain
Your affectionate Son
Julius

Alexandria La May 9th/63

Dear Mother

After the hardest marching our troops have yet seen this place was occupied without a struggle night before last By Gens Dwight & Weitzel. Emorys Division arriving yesterday afternoon , the distance 95 miles, was accomplished in <u>four</u> days marching which under the burning sun was not very pleasant. May 5th order came to start immediately and at 6.30 AM the division was off for the Red river. Thursday was fearfully hot and as the dust all along had been very thick we soon found that this day was to be a little worse than any of the preceding ones. Talk about your dust at home I never knew what it was till lately, let an army pass through a country where it has not rained for weeks and one can form some idea how thick the road must be with the article. Cut up by Artillery cavalry and baggage wagons and then trodden to a fine powder by the thousands of feet course how could it be otherwise than disagreeable to travel in.

That Thursday by good fortune our regt had the right of the brigade so we determined to give the 31st a taste of the beauties of the rear of the division. At night said regt was all used up while our men were comparatively fresh only having traveled 25 miles in ten hours. It is said that the march from Opelousas to Alexandria has been one of the quickest of the war and the weather being so hot made it still harder. Weitzel marched his division 32 miles the last day occupying the City at 11 PM.

I shall write further particulars in a day or two, as I feel decidedly tired to day, and the prospects are now that we shall remain here a week or two. I can give you more time the next letter I write.

My health is very good and I am in good spirits except being tired.

Yours in haste
Julius

Camp near Alexandria La May 11th/63

Dear John

I have determined for a long time to write you a letter but circumstances have prevented, you know just how a fellow feels after a long march and then living in those miserable little shelter tents which have been our abode almost ever since the 1st of March. However, I will try to relate to you what has taken place since writing to Fan presuming you have been posted in everything up to the day we arrived at Opelousas. Drills, reviews or picket duty used up the hours from Sunnci to sunset except when the sun was too hot to be out of ones tens, at such times as these I usually wrote [cross out] my letters. The evening of May 4th a few of us took a ride by moonlight down town and after having passed a pleasant evening arrived in camp pretty late. At 5 the next morning orders came very unexpected and by 6 the division was off for Alexandria; it seems that Farragut sent word to Bank to be at that place by Saturday as fighting was expected to take place both by land and water. Monday afternoon we heard of the great cavalry raid from LaGrange Tenn to Baton Rouge, 800 miles in seventeen days, of course this created a great sensation and we were in hopes to do another "big thing" in whipping the enemy and taking Alexandria. That same day Capt Howard Dwight of Gen Andrews staff was murdered by guerillas. Capt D was Lu Stackpoles chum at college and I think you must know him. His death made his brother, the General, almost crazy and he vows vengeance against all the bands of these rascals. Day before yesterday the General had one shot. Tuesday morning pushed on halting at night after a hot march some twenty odd miles away. Gens Weitzel and Dwight had started in advance of us, but Gen Emory pushed ahead so fast that he nearly caught up with them till the two former marched their commands 32 miles the last day occupying this city without opposition. Col Gooding of the 31st Mass, commanding the 3rd Brigade, does not like the 38th for various reasons, especially since Banks complimented us in a special order, and he takes ever opportunity to annoy or bother us. With but one exception our regt has always been at the end of the brigade on the march whereas in the other brigades each regt changes every day. At last Gen Emory heard of it and the third day out he put us at the head and the 31st at the rear. You ought to have seen the men march, it was neared a doublequick than anything else, they were determined to let the 31st know there was some difference in the marching at the foot than there was at the head of the brigade. That day was one I never shall forget, the sun was burning hot, for you must know the month of May, corresponded with the month of July at home, and as not a drop of rain had fallen for weeks the roads were several inches deep with dust. At times I could not see three feet ahead on this account, but staggered on blinded and nearly suffocated making over twenty five miles between 5 AM & 3 PM resting in all about an hour and a half. The 31st was nowhere, such a set of used up fellows I never before saw, the regt was strung out for a long distance and I think it served as a good lesson for the next day when they took the lead again there was a great difference in their bragging that "they could walk the 38th off their feet." Clothes, hands and faces were all the same color ere the troops had been on the road two hours, but by good luck the route was beside a small muddy bayou so that water was plenty to wash in and as soon as the column halted for the night the first thing was to wash and afterwards use the same water to make coffee. What would Mother say to see me living in this style, she would think that this portion of the Lathrop family had degenerated slightly. I presume you have seen just such times as these so its nothing new. Friday afternoon entered the city and passing along by the river saw the famous iron clad Benton, the ram Switzerland and one or two gunboats. Farragut is below on the Mississippi, Admiral Porter commands the fleet I believe in the Benton. From Opelousas to Alexandria the distance by the route we came is 95 miles and this was accomplished in four days. It is said by our Genls to have been some of the hardest marching ever done, taking everything into consideration. In all since the 10th of March I have footed about 300 miles of Louisiana soil, carrying a greater portion of my 160 land warrant every day on my clothes. Besides sword, haversack and canteen a small breech loading carbine constitutes my <u>harness,</u> and as to shoes to march in I have been about half barefooted from Berwick to Opelousas; at that place I had them repaired so that difficulty was overcame, but had the misfortune to hurt my knee the day before leaving Opelousas, still not once have fallen behind, or left my Company during the whole of this Spring's campaign, and not a step have I been carried, now I am in the best health possible and can stand any amount of hardships. Nary a blister have I had on my feet all owing to those shoes which Newcomb made for my last July. To show you what a rate we traveled I will say that most of the horses are used up and unfit for service until they have somedays rest; any quantity have died, whereas there are not twenty five men in the regt who are on the sick list. The country through which we passed is magnificent, some of the richest I have ever seen, and such fields of corn and cane, how a northern farmer would stare to see them. Friday noon passed by Gov Moore's plantation, which in spite of Banks orders against foraging, the men pillaged. I had the satisfaction of eating for my supper a good chicken which was raised by some of "the old Governor's" contrabands. Poor Gov Moore the Yankees do trouble him a great deal not contented from driving him out of his three Capitals, they have even plundered his plantation and now his cotton and darkies gone, an outcast from Baton Rouge, Opelousas and Alexandria where will he hide his head so that the mudsills shant find him? At Baton Rouge the Essex sent a shrapnel shell into his private room in the State house starting the old fellow out pretty lively. Being on Court martial and having pay rolls to make out I shall not be able to give much time to writing so please say to those at home that for a few days I shall not answer any letters.

Yesterday morning very heavy firing was heard in the direction of Port Hudson and today the news comes that some of Hunters troops and ironclads have arrived at Baton Rouge and the firing is supposed to have been in an attack on Port Hudson. I hope this news is true, but there are so many rumors floating in the air that one can't believe a thing he hears. I trust that the 24th may be at Baton Rouge and should not be surprised if they did come. What is to be done now I cant tell but orders have arrived to prepare to march at an instants notice. It does not seem possible that they intend pushing us on much more as the weather is getting too uncomfortable to be out in the sun. We are expecting transports to arrive and take us to attack Port Hudson, but may go in another direction.

I must now close as other things demand my attention. Give my best love to all and hoping you are entirely well ere this.

I remain
Your affectionate brother
Julius

PS I don't want my letter published especially when they are written in such a hurry as this. Tell Mother that I have received word that my box has arrived safely in New Orleans. Am sorry old boy that you have been sick but it is natural that a man must have his dark days as well as his bright ones. I had mine at Ship Island & am thoroughly acclimated.

In haste

Before Port Hudson May 31st 1863

Dear Mother

It is said that a mail is to leave here tomorrow so I will write a few lines while there is time, but it is impossible for me to enter in the particulars of the siege at present, so you will have to wait until "Port Hudson is taken". For a week the place has been besieged closely, but as yet not as yet not much damage has been done to the enemy beyond a few guns being disabled. During this time the 38th has been under fire more or less nearly every day, although only a few have been hit owing to the nature of the ground giving us every advantage one could wish. I am sorry to report the death of Lt Col Rodman who was instantly killed May 27th by a bullet passing through his heart. We cannot realize that the regt has been deprived of one whom all looked upon with admiration and respect, or that the officers have lost their best friend, one who always granted us every privilege that could be desired. He was sitting down
with the 38th
near Capt Rodman &two or three others when Gen Paine gave him the order to follow ^ the 12th Maine in a charge on the earthworks, and just as he rose to go forward the ball struck him. He died without pain. I heard the shot fired and the "thud" of the bullet but did not know who was killed till a few minutes later, although within a dozen feet of him at the time.

The fighting of May 27th was very hard and the loss on our side was somewhere near 1000. Including a large list of officers. I am safe so far having escaped with only a slight bruize on my shoulder caused by a spent buck shot. I ought not to have mentioned this as you may think it more than what I have written so rest easy as it did not break the skni. For 63 hours our regt was within 100 yards of the works, the advance skirmishers in the ditch, & scarcely an inch of reb could be exposed without a bullet being fired, still away off on our left were a few sharp shooters who have done a great deal of execution hitting many of our men & killing others nearly 500 yards off. Night before last the regt was relieved for 24 hours, when we again took our old position and while I write the crack of the rifle, the whizzing of bullets and the heavy reports of the artillery just in our rear together with the bursting shell are sounds which fill our ears. Gen Banks is waiting for reinforcement & getting up his siege train before making the ground attack on the rebel works. Already have the 30 lb Parrotts dismounted every rebel gun in sight, but this morning the enemy opened from a new battery firing a few heavy shells as soon as the exact position can be ascertained "Johnny Reb" will have to stop his noise in a hurry.

It is said that our regt will be sent to New Orleans as soon as the campaign is over to do provost duty. I hope this is true but it will be too good fortune for us, as it is just the 38th luck to be kicked around everywhere. In three days & three hours we marched 85 miles from Alexandria to Semmesport & in all up to this time have tramped over 400 miles since last March.

Now my dear Mother I must close, promising to write just as soon as the mail leaves, still you must not worry now about letters as I don't have any time to devote to writing. Give my best love to all & kisses to the babies. Remember I am in the hands of "Him who doeth all things well" & whether it is my lot to be hit or escape it is all for the best.

Your in haste
Julius

Before Port Hudson June 23rd/63

My dear Fan

I have before me four letters from you dated Apl 12th 19th May 10th & 24th which have remained unanswered for an age although I wrote a short note the 19th alt. Do not feel slighted if you don't receive letters from me oftener, but situated as we have been for the past fifteen or sixteen weeks it is hard work to keep up our correspondence especially where I have so many to write to without speaking of business letters, descriptive lists &c which are very important to the sick & wounded men in my company. If it is ever my good fortune to be settled in camp once more I will endeavor to make up for former neglects, but at present you will have to "take the will for the deed." I wrote to Joe & Jennie a few days ago giving an account of the assault June 14th, since that time everything has remained pretty quiet with the exception of digging two [saps?] and making arrangements for the last attack which I judge will take place this week. The 3rd Division only number about 2200 men, on the morning of the 14th they had 3000 to take part in the fight, over 800 men were killed or wounded and of this loss there were only a few regts which were fairly engaged, of these the 4th Wis, 8th NH 133rd NY and 38th Mass show the largest proportion. We read of the repulse of the rebs by Sykes Regulars in the Chancellorsville fight and are told that after enduring a tremendous fire for hours they lost 150 killed & wounded; the 4th Wis out of 300 had nearly 80 hit in the fight of May 27th & on the 14th June lost over 130 out of the remainder. Of the 6 companies of our regt in the skirmish line about one half went down and so reduced is the 38th by hard service and hard marching that we have been put in the rear of a battery & shall probably remain there as a support till the place is taken or the siege raised. I hardly think that we shall take place in the coming fight still if necessary can yet do the good fighting which has ever been shown by the regt.

A mail is expected today or tomorrow so I shall probably have some more letters to answer; in the last came the knife from Mother, please thank her for me as it is a most valuable and useful present especially when in such a life as this. I intend to write to her as soon as I have another good chance, but it is of no use now as mails are sent only as far as Springfield Landing this side of Baton Rouge until the Rebel stronghold is taken or the place lost. For the first week or two in Camp I shall have all I can tend to making out pay rolls and quarterly accounts to say nothing of ordnance returns, descriptive lists and final statements of those killed in action. I really dread to think of the awful amount of writing. As soon as we get settled I shall write for a lot of clothes but am afraid it will be cool weather before they get out here, still you can tell Joe when he returns that I should like to have him get for me a pair of pants <u>dark</u> blue flannel with <u>light</u> blue welt, and a vest with the same material, on the vest I want the Mass. State button. I don't care about the finest quality of flannel, but want something thin & cool as the clothes I now have are too thick for summer, John will know the size; by the way I must have the pants made small enough round the waist to wear without suspenders as these articles are quite uncomfortable in this horrid hot weather. If we go to New Orleans I shall need a thin flannel sack coat such as John wore two years ago, but will mail before sending for this till I see where we are to be stationed. My other box still remains in New Orleans I have the receipt yet. Don't let Joe re-enlist again, he may have a good time once, but I rather think the nine months men in this department have had enough of it and I would not have him go through what I have for the world. Just think, the 9th April I left Algiers with 60 good strong men, now all I can muster for duty is 20. 18 have been killed or wounded and the remainder are now in hospital where many will linger for months till death puts them out of misery. Tell him he can have everything of mine to use as his own if he will only stay at home & next winter he can have my skates. I will never forgive him if he goes out again to the war, he may be thankful for escaping this time, the next might tell a different story. If you see Ned Davis give him my best regards. Poor fellow he has had his share of the gory of battle and I do not think he would care to go out again even if his wound would let him. I shall write to him as soon as we are settled quietly again.

I felt rather used up for a few days past but am now alright once more. The fact is we are all played out and need rest after the severe campaign and hard fighting to say nothing of this burning sun which in itself is almost enough to kill one. When niggers are sunstruck you can make up your mind that the weather is [?same?] However, the prospects are now quite encouraging and if this assault is carried out in a decent manner we shall be through with our troubles ere next Saturday night. I see by the papers that the 9th Army Corps is at Vicksburg, they will get enough of it there as that city is fortified by nature & science just like Port Hudson. We have not been paid off since Jan 1st but I managed to get my money to April 1st so after next pay day shall have quite a nice little sum to send home. It looks like rain so I must close as my things out to the battery will get wet unless I hurry up. Lots of love to all & kisses to the young uns. I have received all your letters & contents.

In haste
Julius

P.S. When eating dinner yesterday a bullet came by my head & passed between the dishes, how should you like these little visitors?

Before Port Hudson June 27th 1863

My dear Louise

I received your interesting letter of the 1st May only a few days ago and will now try to answer it, although we are situated so close to the enemys works that bullets are continually flying closer to ones head than is agreeable. The regt, having been badly cut up and reduced by sickness to about two hundred, has been placed in the rear of a battery of twelve guns and two mortars and here I presume we shall stay till the place is taken; our regt has had its share in the glories of the campaign and we have all seen enough service to last for the remainder of our days. Talk of the 44th having seen more service than any other nine months regt. Why they have not done one half the marching or lost one half as many men as the 53rd has. Since the 12th of March our division has marched over 450 miles, fought the enemy five times and is now before this stronghold. The march from Opelousas to Alexandria 95 miles was accomplished in three days and a half, from the latter city to Semmesport a distance of 85 miles we tramped in three days and three hours, and all this you must remember in weather hotter than we ever have North. Old soldiers from the Potomac who were with us say such hardships were never heard of in Virginia and I well know that what I endured in North Carolina was nothing compared with what I have since experienced. At times officers & men were nearly half starved and it was only by foraging on an extensive scale that we could live, the punishment for pillaging is death, but we made up our minds that we were likely to die if things went on in that style so concluded to run the risk. Our fellow was shot and any quantity of officers dismissed the service, but I always escaped by good luck.

I should have enjoyed myself very much at reunion but the stern duties of war keep me from any such pleasures. I was in hoped last year to have been present this year, still one can't very well get a furlough way out here and I hardly think I shall see home again for many a long month unless wounded on the coming assault.

You have doubtless read of the attack made May 27th I which Emorys & Weitzels divisions were engaged, driving the enemy before us and using the bayonet about as much as bullets, the slaughter that day on both sides was terrible, no quarter asked or given. Lt Col Rodman of our regt fell within a few feet of me when about to make our final charge on the enemys works. I was hit in the shoulder by buck shot but only bruised, you can't imagine how mad it made me, so I took out my spite by firing at the rebs with my carbine till I thought I had given one or two rebs a pill. Sunday the 14th inst. a grand assault was made on the earthworks, the 4th Wisconsin & 8th NH formed the first line of skirmishers & six companies each of the 58th & 38th the second. Just at daybreak the charge was made over a wide field, when men were falling "like leaves" before the Autumn gale, down into a ravine, up a hill & down again; here one rested for a few minutes & then made another dash forward till within a few yards of the ditch. Never before did I see bullets fly so thick or men fall so fast, the groans of the wounded and dying still ring in my ears, and this remember was on a bright Sunday morning when instead of attending church we were doing our best to kill our erring brothers. This was my second Sunday fight and I hope we shall never be engaged in another, especially such a one as this. All day long we remained there and when night threw a curtain over the bloody scene the troops withdrew leaving most of our dead & part of the wounded. The 4th Wis & 8th NH lost two thirds of their men and the 38th <u>one</u> <u>half</u> the number of men and officers engaged. The reporters tell some queer stories about the "niggers" in the fight of the 27th May, it may do very well to read, but when it comes to truth I must say that somebody could not have seen that "terrific fighting". The "blackbirds" fought very well, far better than I anticipated but they only lost <u>one</u> <u>fifth</u> of the number engaged whereas white regts. In some parts of that days work, had fully one half killed or wounded. The black troops have shown that they can fight still they did not have half the work to do that our two divisions did, neither were they under such a fearful fire. The white regts feel slighted that so little is said about their deeds and so much about the contrabands, still I am happy to say that Gen Banks considers the 38th one of his best regts of the Corps the 8th NH & 4th Wis, all that these two can now muster are about <u>seventy</u> <u>five</u> men each. I left Berwick City April 9th with 600 good, strong men and all I can now muster for duty are eighteen, one officer and eighteen men have been killed or wounded, the rest are in hospitals broken down by hardships, when a company is reduced in this style you can form some idea what we have been through.

Sunday morning I had to close last night on account of darkness, but will finish this time. For the past two hours the mortar in front of us have been sending their iron compliments to "Johnny Reb", and every few moments a thundering roar tells me that another shell is on its mission of death. It does not seem like our calm, peaceful Sundays at home, the crack of rifles and the louder report of canon, the whistling of bullets and shrieking of shell all show that "there is no Sundays in war". Two weeks ago this morning I was in a place where the missiles of death flew thick and fast, never before in my life did I think my time had come, as when charging on those works nearly all who were with me went down, although my clothes were cut by bullets still I escaped without a scratch. Had our supports only come forward it would not have been a second Balls Bluff affair; our brave Gen Paine fell early in the fight and many a good soldier gave his life for his Country. Three days after Banks sent in a flag of truce and had the bodies buried, strange to say two men were brought out alive having lived all that time without a mouthful to eat or drink; what their suffering were none of us can form the slightest idea, they must have been awful. A number of the dead had their wounds tied up but had died at last for want of care, why Banks did not send in the flag the next day is more than I can tell, it certainly would have saved a number of lives, but what does he care for the poor fellows? Not one of 114 could be recognized, the bodies having turned perfectly black. It was a horrid sight and one seldom witnessed in war. Col Kimball of the 53rd says at Antietam the sight was not equal to this, though at the former place there were a great many more bodies. One large ditch received the remains of these poor fellows and there they will rest quietly till the last day. Of the small force engaged 1500 were killed & wounded; officers are picked off very fast, in some regts scarcely one is left unhurt. Let me now change my "sermon" to scenes of peace.

I can imagine just how dear old Dedham looks this morning, the first bells are ringing and the younger portion of the inhabitants are on their way to Sunday school, while the older ones are getting ready for church. Would that I could once more see the familiar faces at home or visit powder rock "by moonlight alone". I am afraid the time is far distant ere the wanderers return, but hope another year will see us all back again. What excitement there must be at the North on account of the "invasion", the news of which came last night. We are all glad of it as it will arouse the sluggards at home and if Lee is not defeated and cut to pieces this time I fear we shall never do it. I would give a months pay to be in Boston for a few days just to see the militia turn out, now is the time for the returned nine months men to show their patriotism and if they do not go forward I shall think they are a little cowardly.

Mother tells me that Ned Davis is at home, poor fellow he has seen enough of the glories of war and now returns a cripple. John is having a hard time of it and I am afraid he will never be strong enough to rejoin his regiment. How I manage to bear all these hardships is something beyond my comprehension, at times I have been the only Captain in the regiment on duty; not once in all my military experience have I fallen out on a march, though iit has only been by making the greatest exertions that I have kept in my place on some of those fearful marches.

And now Lousie I think you will be tired before reading this long, stupid letter so will not trouble you any more this time. In my next I hope to be able to give you an account of the storming recapture of Port Hudson. Give my regards to all my friends.

Yours truly
Julius

Before Port Hudson June 27th 1863

Dear Mother

I have received two letters form you lately, one dated May 6th & the other June 1st and shall now answer both. Things have remained pretty quiet since the 14th until yesterday afternoon when a heavy battery of ours on the left of the line opened into the 15 gun batter of the rebs. Since the distance from one to the other is nearly 200 yards I imagine the fighting was very bloody but as yet have learned no particulars, this morning they are at it again. As soon as both sides were fairly engaged the gunboats and mortar fleet commenced to shell the works and at times we could see four 13 inch shell bursting at once. Towards night a ten inch mortar in front of us shelled the rebs out of a corner of their works at a place where we made our Sunday attack. At this point a sap has been dug & out men are almost under the works in front of us is another sap which is about completed so we judge the final assault will be made ere many days, probably early next week. I think this time the attack will be successful as it would have been the 14th if the supports had come up, the total loss that day was 1500; three days after the fight Banks sent in a flag of truce & had the killed removed, among the bodies were found two still alive & others who had tied up their own wounds but died for want of care. Just imagine how the poor fellows must have suffered, three days without food or water, & yet Banks could have sent the flag in the next day but would not. 114 bodies were buried that night & none of them could be recognized, they had turned perfectly black.

Our regt has become so reduced by sickness & those fallen in action that I hardly think we shall be put into the fight but support the battery where we are now. I am sorry as we are all pretty well used up and having done our share on the campaign think it is about time we were held in reserve. Gen Paine has lost his leg but it doing well, our division is now commanded by the Col of the 18th NH, Gen Emory being sick in N.O. The rebs have captured Bashear & Berwick Cities & are threatening an attack on New Orleans so Banks will have to hurry up here.

I wrote to Joe some days ago supposing he would be at home by June 20th since then New Sumner wrote me that their time would not be out till July 20th. You can read the letter I wrote him, as it is directed to Dedham & then send it to Newbern. I requested Fan to get Joe to buy me a pair of pants & a vest, the pants will be enough. Have these kept till you hear from me. I also want you to buy for me two or three bottles of Browns Jamaica Ginger, half a pound of crude camphor, three pairs of thin white gloves, get these very small. I had hard work to find any small enough last year, one or two more white shirts, and a pair of French flannel shirts, the latter must be pretty large, those Joe got me last fall shrunk so that I could not wear them. Also I want half a dozen cans of milk, a lot of chocolate & cocoa besides one or two pineapple cheeses & anything else that will be good to eat as well as to keep long enough till I get them. You can pay for all these things & send me the account, I will forward money as soon as we are paid off. I requested Capt Hildrith of our regt who is now at home to buy me a cap & leave it at Mr Wakefields office & gave him $10. to pay for it, the balance he will give to Mr W. You can put this cap in the box: by the way I need three or four large white handkerchiefs, have all my things marked with my name this way. Capt JM Lathrope Co I 38th Reg Mas Vols. Keep the box till I write for it as I may want a thin coat but can't tell till I see where we are to camp this summer. My other box is still safe in New Orleans. I wish I had my net for the musquitoes are very thick here & plague me so that I can hardly sleep.

I am so sorry to hear that John is no better. I was in hopes to have heard by the last mail but Jennie said nothing about him in her last note. Tell John I am extremely obliged to him for those straps, they arrived safely yesterday morning. I think they are very handsome indeed. I am very much obliged to you for the knife & album both are useful presents, the former here &
photographs
the latter at home. Some of these days I intend to have some ^ taken for the officers & will let you have a few. Col Rodman has some pictures in New Orleans which he had taken just before leaving & as soon as we get a chance the officers are going to send for them.

The stories in the papers about the desperate fighting done by the niggers are perfectly absurd, it is true they fought very well even better than I expected, but they didn't come up to the regts under Weitzel including Emorys division, the fighting of our men in the woods was perfectly awful and although engaged in it myself, still had no idea till I went overt the ground again. The baynet & but of guns were freely used on the rebs, it was a regular hand to hand fight for the first line of troops. The nigs had one fifth of their number killed or wounded; in the fight of the 14th the 4th Wis & 8th NH lost two thirds & our regt one half the number actually engaged but as long as we are not niggers we shan't get any praise.

The weather here is perfectly hot I should think it must be about 100° in the shade and as we are in no condition to stand this heat it rather takes hold of a fellow. How I should like some iced claret punch, now & then when near the hospital I manage to get some ice water which I can assure you is a great luxury.

I shall write again to some of you as soon as something of interest turns up, but for a few days I think the two armies will remain quiet excepting of course the sharp shooters at work, one bullet yesterday passed over my table (a cracker box) when I was eating dinner, but such shots as these are very rare. Every now and then our guns give the rebels a little shelling just to keep them from working on their fortifications, but after all do very little damage as the enemy have sat holes to crawl into whenever a shell comes. Yesterday afternoon a ten inch shell from one of the mortars fell on the ramparts & rolled down into one of these holes, as it exploded just then I guess "Johnny Reb" got hurt, the rest of them skedaddled in a hurry. A few days ago a shell blew up a magazine in the rebel works & set the woods on fire, as soon as they commenced to put the fire out several guns opened throwing shrapnel into the crowd. These beautiful moonlight evenings I spend on top of the sand bags of the battery in fron of us, watching the firing going on at different points, or the shelling by the mortar boats.

And now dear Mother good bye, give my best love to all & many kisses to the little ones. I am in pretty good health, though thin, still when there are so few officers on duty I can't afford to be sick. Shouldn't you like some green corn & tomatoes? I have them nearly every day. Will write again Soon.

In haste
Julius

PS If you see Mr Louis Bullard tell him that Lieut Geo B Russell is doing very well although at last accounts the ball had not been extracted.

Yours truly
Julius

5 miles from Port Hudson July 9th

Dear Mother

Port Hudson surrendered yesterday afternoon unconditionally and this morning was occupied by our forces. The news of the capture of Vicksburg was received the 7th and information was immediately sent in to Gen Gardner; early yesterday morning the rebs hung out the white flag and arranged terms of surrender. Late in the afternoon two picked regts from each Division were selected to occupy the place, with the 8th NH our regt started but when near the works on the left centre it was decided to wait till morning so we bivouacked on the road. At 12.30 this AM the 38th was turned out as our brigade had been ordered to relieve Gen Dudley so that the latter might march to the relief of Donaldsonville which the rebs were besieging and thus we lost the pleasure of seeing the inside of Port Hudson, however in a day or two we shall return as the 3rd Division is to garrison the works. Our brigade is now encamped near the scene of Angiers fight of May 23rd and till everything is settled about the prisoners &c it will remain here. I am in great hopes of going to New Orleans soon but can't tell at present. One good thing is our campaign is ended and the Mississippi open. Our bixes will be sent up to us now that Weitzel has started to clear out the guerillas who have fired into nearly every boat which comes up the river from New Orleans.

I recd your letter of the 17th June a few days ago and am glad to hear that John is getting better. Will write again in a day or two as soon as we are fairly settled in camp or when I know what is to be done with us.

In haste
Julius

Baton Rouge La July 13th/63

Dear Mother

Our brigade started from Port Hudson Plains night before last as convoy for a team of artillery and wagons eight miles long, and reached here yesterday morning having marched 19 miles; you can be assured I was decidedly tired and sleepy having been on my feet all night long and if it had not been for the fatigue This place would have seemed a perfect paradise after being absent from all civilization for such an age, as it was we were right glad to be in the city, but how long our stay is to be none of us can tell, we may remain a week or might start at a moments notice for New Orleans.

I want you to send out to me as soon as possible with the other things which I have written from time to time, two night shirts, any of my old ones will do, two pair of small sheets and have Joe or John buy me a thin blue flannel coat such as one as John wore two years ago, have the Mass buttons put on the blouse and I want the inside breast pocket on the lfe side made very strong so that I can carry a revolver in it as it is necessary to have one of these little shooters when going about the streets of New Orleans. Also I want a large pocket on the inside of the coat on the right side and outside pockets John and Joe will know about what I need. Have eyelet holes on the shoulders for shoulder straps the same distance apart as the enclosed piece of paper.

We shall soon be paid off so that I can send on the money, but hurry up the things as I am sadly in want of all especially the thin suit. I write at first for a vest, but afterwards sent a line to Fan that I did not care for it as it is too hot weather here for a vest. send on anything good to eat that you can think of and I will pay for all that you can pile into the box, nail it up very strong and send it as before, taking duplicate receipts; direct it to New Orleans.

The river is clear from one end to the other now and boats pass up and down without molestation, the rebs have all skedaddled from Donaldsonville, but Weitzels is close behind them.

You must remember that things will not keep very well out here so that you will have to be careful what is put in the box. Love to all. Will write again very soon.

In haste
Julius

Donaldsonville La July 17th 1863

My dear Mother

I received this afternoon a letter from Jennie and one from you and shall now commence an answer to the latter. I wrote to you a few days ago from Baton Rouge at which time I supposed we should soon go into summer quarters, but until the country between the Mississippi and Berwicks Bay is once more cleared of the enemy, there will be no rest for the troops. Day before yesterday our brigade was ordered from Baton Rouge to this place and at midnight landed just below Fort Butler, here we have remained in camp in the open field, how long this is to be our resting place I cant tell but think in a day or two the troops will march to Brashear City distant about 40 miles. Last Monday Gen Dudley had a severe fight some three miles from here & was pretty badly whipped, as the enemy had 6000 men to his 1200 since then quite a heavy reinforcement has been sent down the river and any quantity of artillery so that if another fight comes off we shall be a little better off. Magruder is here under Gen Dick Taylor and while we were at Port Hudson they raised mischief around Brashear City and Donaldsonville at one time preventing boats from passing up & down the river. They laid siege to Fort Butler with 3000 men and early one morning made a desperate assault on the work, but by hard fighting its garrison of 150 <u>sick</u> men, aided by a gunboat, succeeded in routing the rebs killing over60 and taking about 120 prisoners. Three of my company were in the fort during the siege and from them I learned the facts above. There are only <u>five</u> officers at present with the regt, the Major, Adj. one Capt & one Lieut besides myself, the others being sick at Baton Rouge, this brings me second in command, but as I am pretty well used up, think if there is any prospect of hard marching that I shall let Capt Hodges act as Lieut Col, and return to Baton Rouge or New Orleans, as I need a little rest. It is no joke I can assure you campaigning in the summer in this hot climate & no one ever expected it could be done, however as soon as these rebs have been dispersed we shall proceed to make ourselves as cool as possible.

In my last I forgot to send you the distance apart that the eyelet holes were to be on my sack coat, and as my baggage is at Baton Rouge you will have to go by those on Johns coat, till Joe or whoever attends to the purchasing of my clothes which I have ordered, to get a good article, but do not care to give an extravagant prices for them, and above all things send them <u>as soon as possible.</u>

I am sorry to hear that John is no better as I was in hopes to hear by this time that he was well enough to walk about, however I presume "it is all for the best" but it seems hard that he should be so afflicted. I am very glad now that you did not give up housekeeping as we all wanted you to. Jenny writes that Billy Bullard has gone to Europe it seems to me that he must have left earlier than was expected unless they have Commencement at Amherst in June, else he did not graduate. Tell Mrs Bullard to give him my love when she writes, I think he might have written me before he left, but as he hated to write letters, presume that it was too much like work.

We all feel badly that our division, which has done the hardest fighting out here, never has a word said about it, in the accounts of the 27th May fight, it was all nigger when in reality the 3rd Div under Weitzel was the only one which commenced the action at the appointed time &suffered the most & in the 14th fight "it was all Grover & Weitzel" when if the truth was told the 3rd, or as it is called here "the fighting division" made the attack having to pass over a wide field before any shelter was reached, while the others crawled up a ravine till close up to the rebs & even then did not get as near the works as we did. We all think that "honor should be given to whom honor is due and considering we have lost so many lives think that Emorys men should receive their full share of praise, but as there is no reporter in this division of course nothing is known of our movements except the long list of killed & wounded. I am so glad the 43rd were not out here some of the nine months regts whose time expired over four weeks ago were rushed into the fight here last Monday, the 48th running as usual that regt & the 4th have disgraced the state, quite a number of the latter with five officers are to be sent to the Tortugas for refusing to do duty after their time had expired. Boats pass up & down from St Louis now the first time for two years & a half, it looks like good times out here, prices of things will be lower. A few days ago news came of the defeat of Lee which event we have been expecting ever since he invaded the North, it now begins to look like seeing the end, if only Gen Meade can cut him up a little more & Dix take Richmond (a rumor to this effect has reached us) I think Jeff will want peace. Sergt Wilson of F Co is now at the St Louis Hospital New Orleans, at last accounts doing very well, although wounded in the head, finger and groin, the latter is the worst wound the others being slight, however as far as I can learn he is recovering as rapidly as can be expected. It is hard work to find out anything about the sick or wounded when we are in the field. I have several men who have been reported dead but cant ascertain whether they are alive or not.

Goodbye now dear Mother, I will write as often as possible. Give my love to all & kisses to the little ones. Don't forget to send all the things I have written for from time to time, You will have to look up all my letters which I have written to you, Fan & Jennie & the last one to Joe to find out what I want. Put any good things in the box which you think will keep & send the bill of the whole to me.

Your affectionate son
Julius

Baton Rouge Aug 20th 1863

My dear Mother

I returned from a very dull visit to New Orleans, and although quite busy today writing up accounts, still must spare a few minutes to write home. We are on the eve of a great movement to Mobile (at least so report goes) and ere many days have passed a large army will be on its way, whether our regt will be included or not I can't tell, the majority of the troops will be from Grants Army and already there are some 15.000 at Carrollton. Gen Weitzel has had his pick of regts to form the 1st Division of the 19th Army Corpts and the 38th is now in the 2nd brigade of that division, from this I think Weitzel will operate with the Western men. Enclosed in a photograph of Gen W. who by the way is a splendid soldier and is almost worshiped by the men, I need not tell you that we all feel highly honored as well as pleased that our regt should be selected by him.

Several of our officers have gone home after conscripts and if you keep a look out in the Journal you will see a notice that Capt Smith or Capt Sewell will take small parcels for the regt so that if there is anything for me it can be given to one of the above officers. ________ I am expecting to receive some letters & my box every day but when I left N.O. night before last the steamer had not got in although then three days overdue, but I presume she will come along all right. It seems as if I had been away from Camp a month instead of only six days, but as the men have been hard at work fixing up the place it shows a great improvement.

New Orleans is filled with officers principally from the 23rd Army Corps, but beyond these everything is remarkably stupid, so that I did not have a good time: the weather is fearfully hot & all we could do was to lie still and drink sherry cobblers or claret punches. Saturday afternoon I attended the presentation to Gen Banks which was a big thing I assure you, the account of which I send by this mail. All officers were present and reserved seats quite near the young ladies & of course this in itself was decidedly pleasant for a fellow like me who had not seen any of the fair sex for so long a time, this being the first visit I have made to the city since the 1st of April. Sunday several of us went to Christ Church, the Union Church of the city, and you can rest assured that I enjoyed it as this is the first time I have been inside an Episcopal church since leaving home, although for months we had that service in camp. But very few time have I had a chance to attend church and always it has as happened that I had no other place to go than a Catholic church, till that Sunday. Quite a crowd of the Union citizens were present as well as a lot of officers who always patronize the church. Tell Joe that I saw Charlie Slack in N.O., he has been quite sick for some time but is now better. I don't think Joe knows him although he is an old friend of mine, but Joe is acquainted with his sister.

My health is tip top except that I am pretty thin as usual especially in the hot weather, the next time I do down to the City I shall have some pictures taken. If you every send out another parcel to me put in it my knife, fork & spoon which I sent home last Winter.

Good by now dear Mother, I am in such a hurry that I can't stop to write more. Love to all

In haste
Julius

Baton Rouge Aug 23rd 1863

Dear Mother

I thought that when I wrote last that it would be very strange if our regt should be left "out in the cold" when others were going to Mobile and sure enough night before last orders were received to be ready to march at a moments notice with three days cooked rations and one hundred rounds of ammunition. So good bye to our long looked for rest which Gen Banks had promised us at Port Hudson; as the 19th Amy Corps with the 23rd are bound on this expedition. I presume without doubt it will be Mobile to be attacked though some are inclined to think that Texas is our destination. I sometime think of going into a nigger regt as they are the only ones who do nothing but garrison duty while we poor fellows have to work hard even in the hottest part of the year. The bible tells us that "there is no peace for the wicked" and I believe it applies pretty well to us, at least one would think so by the way we are licked around in this Department however as the men say, "its all in three years." It makes one provoked, after fixing up my tent in good shape and making up my mind to have a nice time lying idle, to receive such orders and so soon after the other campaign; now I presume when Mobile falls Banks will lead us a nice little tramp through Texas just to keep us in fighting condition, I shouldn't care much if they would wait till cool weather set in, but in this heat I assure you the prospects of campaigning are anything but pleasant, however we shall have to "grin and bear it."

The Steamer has not yet arrived from the South so I am afraid I shall have to let my box lie in the Express office for one while. O am sorry about this as there are things in it I want very much especially the revolver, still if we don't move for a few days it may reach me, but it is very doubtful.

Day before yesterday as I was lying down reading I was startled by the report of a heavy gun and the whizzing of a shell and on running out saw that the gunboat was firing at some guerillas on the opposite bank of the river from our Camp. The rebs skedaddled in a hurry without caring to make a closer acquaintance with the ironclad.

Last night just as I was going to bed I heard an infernal yell coming from the top of my tent and on reconnoitering, discovered a large owl perched on the ridge pole making night hideous with his hootings. I threw a bunch of cartridges at him which frightened the bird away, but early this morning he came back and woke me up with a fearful cry which made my blood run cold to hear. If firing in camp had not been forbidden I should have put my visitor out of the world in a hurry. What with crickets and owls we have music, such as it is all night long; these crickets are the plague of our lives, they get into everything, eat holes in our blankets, clothing and even in shoes, so that we have to keep almost every article of clothing packed. They haven't spare my nice new musquito netting for a dozen hole testify as to their voracity, however they have cleared out of my valise not liking the taste of red pepper which I have sprinkled thickly around in place of camphor.

Enclosed you will find photographs of Col Rodman and Capt Hodges of our regt which I want you to put with my collection. They are both very good, especially Col Rodmans, I think everything of this picture and hope it will only reach you safely.

By the way I want you to knit me four prs of woolen socks as my old ones are all worn out notwithstanding the many darnings I have given them, all I have now are thin ones which are almost worthless in marching. If you get them done in time you can send them out by some of our officers who are now at home, they promised to put a notice in the papers that they would take small parcels for the regiment.

I do wish the mail would come along as we haven't heard from home for two weeks and it seems almost a life time to me, especially as I am looking forward for my box with so much anxiety lest it should arrive too late. I reckon I shall have to go without thin pants this season, still they will do for the next, although by that time I hope to be at home.

Yesterday I finished the last of my accounts and as they are now on the way to the War Department I feel very much relieved. Govt expects us to keep our accounts in good order and to send them on, when we are in the field, just as promptly as if we were in camp. What we shall do now I can't see as we are not allowed to carry anything except what we can take in our hands, so I shall let my book and papers slide for one while till we are in a settled place once more.

I forgot to tell you that Capt Hildreth, whose photograph I enclose, lost his note book & with it Mr Wakefields address so he returned me the money. Hildreth was one of the best officers and when in the 1st Mass, Regt passed safely through over a dozen different battles. Poor fellow we had to part with him as an injury he received in his old regt came on again and he resigned being unable to march.

Down town there are all sorts of rumors as to our destination, some say we are to make a junction with 9th Corps & cut off Johnston, others that we are to attack Mobile, the latter I think is the right supposition.

I believe I have mentioned in my former letters that the 23rd Corps is at Carrollton. I was mistaken in the number, it turns out to be the Army of the Frontier under Gen Ord, the 13th

Corps. With such old soldiers as these we can't help winning, and wo be unto the rebs we come across unless they greatly outnumber us. I hardly believe we shall all get away under a week, Gen Franklin, who commands the 19th Corps has sent up word to have all the white regts march from Port Hudson & they are expected here to night. The "nigs" are the only ones to be left, they are even taking the white men from the forts down the river & putting blacks in their places, so as to make the old troops take to the field, every man they can possibly raise is to go, but I hardly relish the idea of going into action with such a few officers only ten at the most in our regt & some 300 men can start, other regts are in the same condition, however theres all the more chance of winning glory.

And now dear Mother good bye till we are off, then I will write a few lines. Love to all & many kisses to the babies.

In haste
Julius

(Letter fragment)

Camp William L Rodman Baton Rouge Sept 30th /63

My dear Mother

I received your letter of the 31st Aug. just two weeks ago and should have answered it sooner had I not thought you would be at Bordentown, however "better late than never."

At the time of writing my last letter to Joe, an attack was expected on this place and everything was in readiness to meet the enemy, whom it was supposed would make theur appearance that night as the wires had been cut above and below about the same time of day. Large and small guns were placed on all the roads leading into town and all the works stationed inside the rifle pits. Our regt was reinforced by some forty convalescents from a dozen different regts and as we were on the road where the fighting was expected to commence all of us officers has the most valuable things we possessed packed up so that our niggers could save them if we were forced to fall back to the open field. However all danger is now over, as Gen Orsne with a brigade from N.O. kanded below here and is now scouring the country in our ready. The 4th Wis Cavalry have captured quite a number of the guerillas and I trust dispersed all the bands. Day before yesterday all the guns were removed to the fortifications and our picket line reduced to its former dimensions. After all it was a big scare, and although Col Gooding was perfectly sure we should be attacked, still I do not think any of us felt there was any danger, yet we were all prepared so as not to be caught napping. The firing I wrote Joe about I afterwards learned was from the Essex and Fort Butler shelling some guerillas near Donaldsonville. We can't learn anything about the army at Brashear City although there are all sorts of reports some say the movement is a failure, others that it is only intended as a blind while a dash is made in some other direction. I am inclined to be of the latter opinion, but can't make up my mind till I can see someone who knows <u>just</u> where the troops are.

Yesterday and to day we have been blessed with a real hard storm, just such as I used to like at home to drive in the plover and ducks, although there it invariably ended on a Sunday so that the birds were off by the next morning. We have not had any rain for weeks and everything is parched up but now the ground is pretty week soaked much to my delight. Last Monday afternoon a Mr Briggs, Asst. Engineer on the "tin clad" Signal, came up to see me as he heard I was from Dedham. I soon found out that he is a brother of Richard Briggs, successor to Summers & Co and is well acquainted with the family of S's He went out in the 13th Regt with Bill Darnell and several other of the Dedham boys, but was afterwards transferred to the Navy. I went on board the boat and took Supper with him and had quite a pleasant time. His boat cruises between this place and the Red River so that we can see each other occasionally. I can assure you it's quite a treat to meet somebody who knows any of ones friends in Dedham.

For the past two days I have been very busy making out my clothing and Ordnance returns for Washington and this morning finished the last. I received a letter from Joe and one from Jennie yesterday with several papers so now for a few days shall have something to read, especially as Molly sent out six or eight N.Y. Heralds.

Did you ever get the $10. I sent on to you about the 16th of this month? If so let me know when you write again. Mollie wrote that John enclosed a photograph of hers to me in one of his letters. I have never seen it and am afraid the note must have been lost or else he did not send the picture. Tell Joe that he need not send me my bowie knife, as the prospects are that we shall not have any active service to do this Winter but remain here to garrison this post. I hope so at least, but army life is very uncertain. (END)

Camp Wm L Rodman Baton Rouge Nov 12th /63

My dear Mother

Received your letter of the 26th night before last and was very glad to hear from you. The flag you spoke of has been presented to the regt. And I can assure you it was a splendid one, some who pretend to know say it is the handsomest one in the Department.

You need not be afraid of my doing anything against Mr. F if he has taken the oath, which he said he never should. If I can ever get to New Orleans I shall crow over them about it, I thought he was one of those who would never give in, but was to fight till every cent of his money was gone and then commence life over again. I find the "Almighty Dollar" is king here as at the North, and to keep it a Southerner will forfeit his word in spite of what is said of the "chivalry."

I wrote John yesterday that the cavalry had a little skirmish a few miles out since then I have learned the particulars. It seems that a forage train guarded by two companies of cavalry and one of infantry was sent out on the Port Hudson road. When just beyond Montecino Bayou the rebel pickets were seen and driven in some distance; suddenly the men came on quite a large force of rebels and at nearly the same time discovered they were flanked. The
at the
cavalry dismounted and commenced the skirmish;^ same time sending one back for reinforcements. Our men held their ground in spite of being outnumbered and gave the rebs a warm reception. I have not yet heard what the results were, but suppose the rebs skedaddled as soon as the regt was brought up. All the wagons I believe were saved. It needs a brigade of mounted infantry to keep this country clear of the gangs of bushwackers and guerillas, the latter go round dressed in our uniform and commit all kinds of barbarities; a few days ago they set a house on fire in which a little boy was burnt to death, and this because his mother was bringing in cotton! I tell you these poor families outside our lines have to suffer from both sides, they have their share of the rebellion and a bitter pill it is too. Perhaps they would like a Southern Confederacy instead of the old government. They are sick enough of the war and would be glad to have peace on any terms, what they would do if the rebs held this place I don't see but some of the wealthiest came mighty near starving last fall. On the whole I think most of them are coming down a little as sesech is about played out. The girls think they might was well get acquainted with officers, for there are very few outsiders and they have tried hard to hold back,

but it's no go. Crowds are daily taking the oath still for all that I wouldn't trust one of them in spite of all their Unionism.

I am very sorry to hear of old Mr Rands death, they must miss him very much _________ What lots of changes there are in Dedham. I shall never know the place when I return, all my old friends will be married off, or moved away and I shall be compelled to go with the generation now springing up. Dools mansion has gone they tell me and I see by the papers that Coolidge's block is to be sold and torn down. I hope they will put up some nice stores and a large hall.

What kind of weather are you having? Cold I suppose. Wouldn't you like to be out here now, the days are so warm that I have to keep my tent open and the nights cold enough to make one sleep sound. I want John to come out and pass the winter with me. I shall have a building to live in as soon as I go to Plaquemine, but when that will be is more than I can tell, but I hope soon. Thanksgiving I am going to have a bully good dinner and lots of egg nog, last year we dined on board the Baltic and had a mighty pleasant time.

I hope my box has been sent by this time as I am very much in need of the things I wrote for, especially the Cap. ____ What did Capt Fletcher have to say about our campaign? I presume he gave you a long account of our life in the rifle pits before Port Hudson and the battles. I found out one day that several years ago he used to visit the Bullards, on which occasions Billy and I always cleared off out of sight because he was a city boy. It's queer how persons turn up and speaking of that puts me in mind that I have got acquainted with the Adjt. of the 128th N.Y. who is an intimate friend of Henry Howards. He has lived at the same house with Henry ever since he went to Poughkeepsie.

I shan't be able to write but a few lines more as its about time I was starting for Court Martial. This is a mighty lazy life, I had much rather be with the Company at Plaquemine, though I am not sure I shall have a thing to do there, as all the men are either on picket or guard the whole time. However it seems like home to be near the Company, besides it is a very independent position, only one officer in the place over me and that a major. I anticipate having some jolly times, but may be disappointed.

And now I must say goodbye, give my love to all and many kisses to the little ones. My health was never better than at present.

You affectionate Son
Julius

(Letter fragment)

Plaquemine La Nov 25th /63

My dear Mother

I received your letter of the 12th last night. You see by the heading of this, that at last I am down here having a good time. Monday noon I arrived and instead of only the three or four companies which first occupied the place, I found parts of four Western regts and two Ohio batteries. It seems that they had a big scare in New Orleans and sent this force up here in expectation of an attack on us. As there is a very large body of rebs at Morganza just above Port Hudson. They fire into every boat passing and as the water is less the gunboats can't hurt the enemy on account of their being sheltered by the levee. The Western fellows are a wild crowd and I hope they will be sent away very soon, as our command were having things their own way before the others came. Yesterday morning Col. Sheldon, Commndg the Post sent our detachment out scouting. We started in a heavy shower and as it had rained all day before the roads were in shocking condition. We all arrived at Indian Village a distance of ten miles in a little less than three hours; here the artillery and infantry rested while the cavalry with Maj Bache crossed the Bayou and advanced. The soon came across four rebs and chased them about six miles but without success and on their return stopped at the rebs quarters and eat their dinner for them. We returned to Plaquemine about 4 o'clock and reached our quarters in two and one half hours. Pretty good marching considering that the men have not had a tramp for such an age, and this morn I don't feel it a particle. I should not know we had been on a march if it wasn't for the sight of my wet clothes.

The Company is in the Court House and are nicely settled for the winter if they only leave us here, which I am afraid they will not, things are too good to last long. A lieut. Who is detached from another Co. to mine, is in a small house with me; we are as pleasantly settled as one could wish for in the Army, two large rooms, chairs, tables &c and last, but not least an open fire place. As I said before we are too comfortable and I should not be surprised if we were ordered back to Baton Rouge or some other mean place, although I had pretty good times at B.R. still this is decidedly the spot I have been in yet. ____ Tomorrow all the officers of our command are to have a big Thanksgiving dinner and as there are about a dozen of us we shall have a jolly time of it, although of course cant have such a dinner as we had on board the Baltic last year. At present all the work our men have to do is building a fort which is now nearly finished, the "Greeks" as we call the Western chaps, do all the picket &c.

I sent in by Lt Spear, formerly my 1st Lieut, but was transferred to the Invalid corps, a small bundle addressed to Mr W. containing a little testament and a pair of thick drawers. Give Joe the latter as I never want to see them again, they are too thick for me, besides I have plenty already. I will send the other pair on as soon as an opportunity offers.____ The gingerbread and crackers were mighty good, some of the latter served for my rations yesterday. I will write Aunt Jane just as soon as I find time, but at present I am as busy as possible making up the everlasting returns and writing in the Co Books. Having been away from them for so long everything has got behind, however by the first of next month I shall be all right again.

How is it that Arny Round is at home now? It seems to me that he has a mighty long furlough. Enclosed you will find a picture of Capt Wyman of our regt and Asst. Surgeon Thompson. I have had some pictures taken but they have not yet been sent down, I expect them (END)

(Letter fragment)

(1)

Merry Christmas and Happy New Year to you all.

Plaquemine La Dec 9th 1863

My dear Louise

Several days ago I received your most welcome letter of the 15th ult. and can assure you it caused me "to rejoice greatly." It is a perfect age since I last heard from you when we were at Port Hudson I think, two weeks after our fight of the 14th June. Then I was sitting under a slight shelter with the enemies bullets whistling over, or around me and now and then a dead or wounded soldier would be carried by. Now I am nicely quartered in a small house where my 1st Lieut and I live, everything around me is comfortable, a bright fire is burning in the grates as the day is rather cool, and were it not for my "trappings" hanging on the wall or kicking about I might almost fancy myself at home. Such you see are the changes in a soldiers life, one day in the midst of danger and the next living in the happiest style imaginable.

I infer from your letter that you have enjoyed yourself hugely this Summer, that's right "go it while you're young." I believe in that maxim although some "old maids" think a young lady has no right to have "a time"; some persons I could name in Dedham would get up a scandal if they saw one of the beautiful damsels, for which that town is famous, flirting with one of the other sex. Don't mind them Louise "go in and win" but don't marry one of those "Britishers" even if they make you a dozen offers. By the way I expect to hear of your engagement as most of the girls at home will be married off by the time the soldiers return. They tell me that the North was never so gay as it is this Winter; it really provokes me to think that the majority of the people there do not realize, or care that there is a war going on. Sometimes I think I am losing the best part of my life out here in the Army, but second thoughts tell me that those years spent in fighting for the old flag will never be regretted. You ask me "if I am as zealous as ever in the good cause," all I can say is while life and health last I am in for fighting and hope I shall be spared to lay many a reb as low in the dust as some I have already done. I am sorry to say that the young men now in the North are too fond of pleasure or making money, that instead of trying to do their share of putting down the rebellion, they are staying at home enjoying themselves. You may say "where are my two borthers? I only wish one of them was in the Army once more. Joe I want to stay with Mother to look after things at home. John, I am sorry to hear, has been obliged to resign on account of his long sickness. It was a great disappointment to him as well as to myself that his career as a soldier should have been closed so soon and just at a time too when it looked the brightest. I don't think anyone can reproach us three for not being patriotic, if you wish to call it by that name, still I only regret that I must be the solitary one. The young fellow in Co. D 43rd Regt, who refused to go to the front after their time had expired ought to be kicked out of the town. I never wish to see one of them. Poor, miserable cowards, soldiers is too good a name to give them, a man who fails to do his duty just because his time has expired, is not fit to live with white people, he ought to be sent to some desert island there to remain for life. If I was a young lady I would not associate with one of the skulkers, it is only setting a bad example to others. If a soldier knew his friends at home would not speak to him unless he performed all his duties faithfully, it would do a great deal of good, but when one sees the cowards treated as well as others, it is enough to discourage anybody from trying to get ahead. I admire the spirit most of the Southern girls show they refuse to speak to a "stay-at-home" on either side and sooner than associating with one of that class choose to keep Federal company. At Baton Rouge any pleasant day you can see crowds of fair damsels out walking, or riding with officers, and all of these girls have brothers or fathers in the rebel army. The fact is most of them have got over their hatred of the "blue coats" and are glad enough to have the Yankees come to see them. I have made some very pleasant acquaintances out here among the sesech especially at the planters house and many a good time have I had this fall with persons who six months ago would have scorned the idea of associating with a Northerner.

Joe can tell you all about my wanderings since the surrender of Port Hudson, that is if you feel sufficiently interested in an old friend's welfare to enquire after him. Of the time at Donaldsonville just after the second fight when I lived for days on nothing but green corn, once in a while "scaring up" a hard bread, only one line officer, besides myself, and one hundred and fifty men for duty in the regiment, at a time too, when the sun was hot enough to roast a person and nothing for us to take shelter under but ragged pieces of shelter tent. I tell you what it is I never was and never expected to be so reduced for food and clothing as I was those last weeks of our campaign. I had no money, not a cent to buy rations with, the commissary's department would not trust, although Gov. was owing us six months pay, over $750., and all I had to eat was what I stole from a cornfield close by. My uniform was in rags, my shoes full of holes, so that I was actually ashamed to show myself anywhere, but the last day drew near, the pay master brought us up our greenbacks and at last we landed in Baton Rouge where our baggage, which had been stored since the first of April, was sent and a few days after I was a different person. Such are some of the pleasures of a soldiers life. Just imagine me living in that was at home, obliged to steal for a living and then after the corn was boiled I used a shingle for a plate, all of my eating utensils having been left behind by a mistake of my nigger. Don't you wish you were a "bold soldier boy"? after all it does one good to experience the vicissitudes of life.

From the first of August till the latter part of October, my Company remained quietly in camp with the regiment, with nothing to vary the monotony of our quiet life except now and then the report that the rebs were coming to attack us. By good luck I was detailed on a Court Martial early in October and then had a very easy time no drills or Field Officer of the day to bother me. A little over five weeks ago my Co. with one from the 128th N.Y. one from the 156th N.Y. Co A 4th Wis. Cav and a section of the 18th N.Y. battery, all under the command of Maj Bache 31st Mass came down the river to this place, some twenty miles from Baton Rouge. Here everything went lovely till Gen Stone getting frightened in New Orleans, lest it should be gobbled, sent up a portion of the 3rd brigade 1st Div. 13th Corps. This of course spoilt all our fun, for these Western fellows have done so much damage since their arrival that persons once friendly to us have now become disgusted with the troops. However ere many days one portion will have to leave and I am afraid we shall be the ones. Plaquemine is a very pretty little town and is a spot where a fellow can enjoy himself to his heart's content. My Co. is quartered in the Court House, just across the street from my "Hotel," and almost a week ago we were highly honored by being made mounted infantry, so that now I have a festive steed to take me round wherever I choose to go. Two companies were mounted for the purpose of making a raid into the interior where there is a small force of rebs and with the cavalry can turn out a very effective fighting battalion. Perhaps after all our drills in Cavalry (END)

(1)

Hd Qrs Detachment US Forces
Plaquemine La Dec 12th 1863

My dear Fan

Two or three days ago I received a letter from you of the 26th also a paper of that date and one of the 27th. The papers you send Thursdays always arrive safely, I have already written about my Thanksgiving dinner so won't repeat it, you can rest assured that I had about as jolly at time as one could well expect away from home. You ask me what is the object of our troops occupying this town, no reason in the world that I can see unless there may have been more or less contraband trade into the interior, and now that we are here it opens up a great deal of cotton into our lines. How long we are to remain here is a matter of uncertainty I should not be surprised if we were ordered off any day. At present I am in command of the detachment formerly under Maj Bache, but he being relieved to rejoin his regt. Gen Cooke ordered me to take charge of the battalion for the present or until he could send a field officer down. My Head Quarters are in a large three story house close by the river so that from where I am now sitting I can see up and down for long distance. It is a very pleasant spot, but how long I am to enjoy it is more than any of us can tell. There is a report in Baton Rouge that our regt is to be made into mounted infantry and sent to Texas, this is good news for I like the idea of having a horse to carry me round, however should much prefer the life I am now leading to any I have yet had in the Army and as long as we are stationed here shall not care to go elsewhere except home.

You need not being afraid of my being gobbled by the rebs when out visiting. I know the country thoroughly and as long as the outpost remains where it is I am as safe as at home. Every day I ride outside the picket lines here and yet not a reb is seen this side of the Indian Village, nine or ten miles away; the fact is there are no small bands of the enemy prowling around the river except above Port Hudson, it is a different country than in Virginia and pretty well cleaned out of every reb. Besides the cavalry scout in every direction each day and with the exception of in the interior, never see an enemy. I speak of the frequent scares we have, I never saw anything equal to it, but then the safest way is always to be ready, now last Sunday some 6000 or 8000 rebs were to have attacked this place early in the morning but finding from their spies that we were strongly defended and the men all ready for a fight, concluded that it wouldn't pay and so march back. I have orders to burn one portion of the town as soon as the first shot is fired to give the guns of our fort a chance to play to effect and it only needs the appearance of the enemy to lay those buildings in ashes. Where I now am in a safe place, but I shall have to command the right flank and in case we are hard pressed fall back into the works, as it is now, we have the levee along the Bayou to fight behind, but I don't believe the rebs are any such fools as to think of coming in.

We glorious news we are having, no sooner do we get over the victory of Grant over Bragg than comes the good reports from Burnside. One day last week we heard we heard that he had been obliged to surrender to Longstreet, this I didn't believe for I well knew my old commander would not be caught in that style. Things look decidedly bright and if they continue in that way without any reverses we may expect that the war will soon be over.

We are having glorious weather now, as warm as August weather at home, just the kind I like for out doors exercise, riding or shooting. Since I have had a horse I am continually on the go, so that I don't feel particularly like being cooped up in a house writing. Still I must answer letters as they are Godsends to us poor fellows.

I should like to look in on you all, but now that I have been away for such an age I have become accustomed to this life and do not miss home comforts as much as I used to. You needn't think I have forgotten any of you for that is out of the question, whenever I am having a good time I always wish that some of you were with me to enjoy it. I only trust ere another year passes that I shall be permitted to see old Dedham once more and all those I have left over two years. Do you know it is just over two year ago the 9th of this month that I first left in the 24th? I celebrated that anniversary by assuming command of our battalion here, when on the first occasion was nothing but a Corporal.

Col Sheldon & his men are still here, so now we have some 1300 to defend the place. There are all kinds of rumors afloat, that my men are to be returned to Baton Rouge and the Western boys left, or that they are to rejoin their brigade and I remain. This I would like muchly, especially as the rebs in a few weeks will not be able to attack us as the water is rising so fast they cant escape across the Atchafalaya & the gunboats once in their rear they are gone.

(2.)

Sunday afternoon I had to leave off last night as supper was ready so will finish this afternoon. To day has been very warm and just as church was out a violent thunder storm came up with hail, now the sun is out bright again, I only hope it will keep so. Capt Rodman came down a few days ago to muster several officers; last week he went to Port Hudson and removed the Col's body to send it home. When the coffin was opened they found the features had scarcely changed and the corpse not a particle decayed although on the day he fell the body remained some hours in the hot sun before it was carried to the rear.

It now appears that the force of rebs who were threatening Baton Rouge, made an attack on Natchez but were repulsed; if they can find any of our troops unprepared they will always pitch in, now if we had not been ready last Sunday, "Johnny Reb" would have made a raid on us.

Monday evening You see I write little by little as often as an opportunity offers. This morning I was off scouting and on my return found a Paymaster had arrived with the greenbacks, much to our joy, as soon as some of the officers go down to the city I shall try to get a check from this Paymaster on New York as it is safer than sending by Express besides not costing anything. Everything remains lovely down in this part of Dixie, except that the roads were decidedly muddy this morning on account of a heavy thunder shower we had yesterday. I have not yet been able to visit New Orleans for one reason or other, but shall now try to go down after Christmas.

As I have written you a pretty long letter I shall now say "goodbye". Love to all.

Your affectionate brother

Julius

PS. I have scribbled off a few lines to Johnnie. I can't talk to him as I once could having got out of practice, now, however I think the little one will be satisfied.

Baton Rouge La Jan 2nd 1864

[Note: A "3" in the date was replaced with a "4" (1864) in a handwriting that is different than Julius' and appears to be in pencil.]

My dear John

I am commencing an answer to yours of Aug 23rd which only reached me a fortnight ago, having been directed to your own regt as I have before written to Joe. To night I am feeling decidedly stupid, owing to a cold which I took some days ago and as the weather outside is more like what you are having at home, the prospects are not favorable for its leaving me. Yesterday and to day are the coldest I have seen since leaving home over two years ago; ice formed last night two inches thick and unless it moderates in a day or two we shall have everything frozen up. Such weather as this surprises everyone here, for none can remember such a winter, how should you like living in a shelter tent? Last New Years we landed in Carrollton and that day was a very warm one. I will remember how strange it seemed to me to see ladies out walking the 1st of January with nothing on their heads and carrying a parasol to keep the sun off. I reckon there's some difference this year. My tent is fixed up in gay shape, better than I have ever had before, a nice door keeps the cold out and as the sides and ends are boarded up to the top of the walls, I can keep nice & warm with the aid of a large brick fire place opposite the door. The fire, these cold nights looks jolly, far better than the little stove I used to have. I can sit here for hours at a time in my lounging chair, with my feet on the mantelpiece, looking into the fire and thinking of the past, or wondering about the future. You ought to see what a jolly bed I have made of a tick filled with Spanish moss, and having sheets and four thick blankets, can sleep as sound as a bear in his den. A large flag hangs over the looking glass and above that is my little carbine. Having a tent framed is a great improvement I can assure you, no storm can blow it down or shake it in the least, and with a little kerosene lamp I can sit here and write, no matter how hard it blows. For a carpet I have an old condemned tent, this prevents the wind from coming up through the cracks in the floor. I'll bet now I am more comfortly situated than you are and enjoy it as much as you do your room, even if you have three large bookcases. There's nothing like soldiering to teach a fellow how to look out for himself and I think when I get home I shall introduce some improvements in my room. Don't you think I am nicely settled?

I believe I wrote Joe and account of my doings Christmas so will not repeat. New Years eve I sat up with the Adjt. Of our detachment at Plaquemine, and amused ourselves talking of old times, every now and then taking a drink of egg nog, the favorite drink here at the holidays. He was very well acquainted with Gen Ferrero, who I see by the papers defeated the rebs at Fort Sanders. I always had a pretty high opinion of that man's fight abilities from what he did at

Roanoke and Newbern. At the latter palce I got slightly acquainted with him one day, and ever after he always stopped to speak with me whenever we chanced to meet. Yesterday and today I have kept close in my tent, as it is too cold outside to move round much. Last evening I mustered up courage to run down town & make a call on a lady where I met a whole crowd of people and we had a very pleasant time playing euchre, the eternal egg nog was brought in, as usual out here during the season.

Hurrah! It rains, no inspection I hope tomorrow; if it's anything I hate it is Sunday morning inspection, somehow or other I always detested it. One thing is sure this rain will change the weather and the cold snap will be over in a few hours, by midnight at the latest. Bully for that.

What kind of times are you having at home this Winter? I presume it is the same old story, coming down to breakfast at 7.35 & then rushing for the 8 o'clock train. Does mother bring up your kettle of hot water every morning, or has camp life broken you of the old habit? Evenings I presume are taken up reading, writing and once in a while making calls, or perhaps studying our some chess problem. I fancy I can see you and Mother to night sitting before the fire, she reading the paper and you with your feet on top of the stove reading some military work, or the last Atlantic. Joe of course is running round after some girl or other and at this time (7.15) is putting on his boots to start. What wouldn't I give to drop in on you all just now! Well there's no help but to "grin and bear it" for nearly nineteen months more.

Why can't you talk some of those loafers round town into shame for not enlisting, I'd just like to get my hands on some of the cowards out here, I would make them step round lively. It does make me so made when I think of the way so many of the town have acted, won't go unless they can have a commission. Ned Davis writes me every few weeks and I see by his letters that he is down on the stay-at-homes as much as I am. He said he enjoyed life in the Army although his time was short and in spite of the fearful suffering he has been through, only regrets that he can't go back. I wish all the young fellows in the town were like him, the quota would have been filled ere this. I hear that there is a recruit for my Co. in camp at Readville though what his name is I haven't been able to learn. At the present time I am the only officer on my co. pay rolls, no first or second lieut. From the time the regt. left the State this Co. has had eight officers, several have been transferred, one died and the rest promoted leaving me alone in my glory, however I have a 1st Lieut detached from another Co. to help me.

Did you see our regt. Flag at the Sanitary Fair? I hardly think it reached home in time to be placed with the others. It shows pretty hard wear and a few bullet holes, but as a general thing our color bearer was too great a coward to put it where it could be hit much. I think I should have shot him, if I had been near enough to, once when he left the flag lying on the ground in the charge on Port Hudson, for that he has been reduced & now we have a fine fellow in his place. I never had but one man run in my Co. and that was the 27th May when advancing, I missed him a moment after, but the fellow had got out of my sight. On the 14th June I made him go by my side and he saw a cocked revolver in my hand ready to lay him out if he started on the back track. I only left him when a rebel bullet passed through his leg, since then I have heard that he was tried for robbing letters at the hospital in N.O. & sent to the Dry Tortugas.

I shall write to Fan tomorrow, so goodbye for the present. Love to all.

Much obliged for that picture.

Your affectionate brother
Jule

Baton Rouge Jan 14th 1864

My dear Mother

I received your letter of Dec 4th this afternoon, Sergt Wilson having just returned from home. It seems that he had to wait in New York about three weeks for a transport, and this accounts for the delay. I am now looking for a mail and my valise, which I presume must be in N.O. by this time, but it's no use for me to wait writing if I am going to send anything by Saturday's boat. I was in hopes to be able to tell you it had reached me safely this mail, but it hasn't got up here yet.

The weather today has been very warm and pleasant, the first one we have had since the day after Christmas. It now seems something like Spring and if the mud dries up we shall all rejoice at the change. For the past few days Gen Cooke has had one of his periodical scares, tow mornings ago we were called out at 5 AM and stood in line of battle several hours, and this afternoon we had to go into the rifle pits, each man having 100 rounds of ammunition. It seems that Lt Earle of the 4th Wis. Started out on a scout Monday with seventeen men; some twenty odd miles from here he came across a rebel camp, charged into it and captured about twenty five prisoners, out of three times his number, with all their horses, and made his way out with only one or two hurt. When swimming the Comite he was set upon by a large force on both sides of the river, but notwithstanding the odds against him, he pitched in and had a fearful fight, six of his men managed to cut their way through and brought the news. They say that Earle fought like a demon, after emptying his revolvers he went in with his sabre and killed
he hit
or wounded <u>six</u> <u>rebs</u> with that alone, one fellow ^ had his head cut completely in two parts and the others had most awful gashes, none will get over it. Earle is a very powerful fellow and one of the most daring men I have ever seen, he never knew what fear was, but this time it is all over with him I am afraid, as he was too well known to escape and having a high price set on his head is without doubt dead. The rebs were more afraid of him than all the rest of the troops here put together. He was the one I have so often written about in my letters as being promoted for bravery at Port Hudson. How many rebs he has killed I can't tell, but the number is enormous and being a dead shot he seldom missed his mark. I saw him only a few days ago and was talking with him about the enemy trying to capture him; he said they would have a tough time of it trying it on and I reckon they did, even then he would not have been lost had not his horse been shot down. He also told me that he never should be taken alive. I am afraid his words have proved true. Thus we have lost the best scout and the most dashing cavalry officer in the Department; to show you what kind of fellow he was I will tell you of one or two affairs in which he was the principal character. He has captured nearly every officer, including

the Col, of one of the reb cavalry regts, besides any quantity of the men, and almost every time he went out, somehow or other he always brought in as many prisoners as he had men. One day the rebs set a trap for him, but he sprung it on them and captured everyone in the crowd without a shot being fired. I think the rebs will now become bolder, as long as he is out of the way for we haven't another to take his place.

It is reported that Gen Wist Adams with 2000 cavalry is outside intending to make a dash into this place. I only hope he will try it on. I reckon he will be whipped as badly as he was a Natchez, but you can rest assured no reb will ever attack us, there are too many 30 pd Parrotts in the fortifications and they have a wholesome dread of those guns.

Last night I went to a dance got up by a few officers; there were only about a dozen couples so we had a jolly time I can assure you. The ladies were from the first families in the town and all but one were strangers to me, however now I have a lot more places to call, so don't think my time will hang very heavily in my hands if I choose to avail myself of the invitations to call. This is the first time I have danced since entering the Army, although I have been to lots of parties where it was going on. We had the fine band of the 10th NY to furnish music and I can assure you the whole affair was a perfect success, the crowd not breaking up till near three o'clock this morning. I only hope I shall have a few more such times this season.

My health is bully, although I have a slight cold like the rest of the soldiers, weather as we have had lately is enough to make anyone sick.

Love to all and kisses to the babies.

Have written about the money.

Your affectionate Son
Julius

Baton Rouge La Jan 17th/64

My dear John

Yours of the 1st came to hand yesterday morning. The papers have been duly signed and witnessed and will go back with this letter. I trust now everything will go on strait and that you will have no bother about the money. I have already written how I wish it invested, $100. To be kept in the bank all the time and the rest in U.S. Bonds.

Everything is quiet here today although we still have to turn out before daylight every morning, remaining under arms and hour or two, yet the danger of a raid is now passed. It seems that Wirt Adams, with between two and three thousand men and some artillery, was to have attacked us a few days ago, but his horses gave out before the guns could be brought near the town, the roads being in a fearful condition. Thursday afternoon, I think, we had to turn out and go into the rifle pits, each man with 100 rounds cartridges as information was received that Adams was only a few miles away, but he didn't come. Friday, two Ky. Regts and the 2nd Ill. Battery came up from Plaquemine; with these reinforcements and three gunboats, the Conestoga, Genl. Price and Kenwood, I think we could everlastingly clean out about any force the rebs have around here. I only wish they would try it on. I never believed the enemy ever intended to make a raid on us, but this time there seems to be but little doubt that such were their intentions, however they have gone back now.

I am detailed again on a General Court Martial to assemble tomorrow. Col Chickering is the President and Maj. Drake A.A.G. is the Judge Advocate. I don't think it will last many days although the first case is a very important one and as the Court was called together by Banks it is most likely that special case will be all. I hope not however as I don't like being officer of the day once a week and having to ride 30 miles in such weather as we have lately had, to say nothing of rising before daybreak & standing out there in the cold. For the past few days I have been in command of the regt, all my senior officers being on detached service and now that I am on this Court Martial the charge of camp will go down lower than it ever has before, to the ninth Caapt.

I am very sorry to hear that you are at last a citizen, for I was in hopes that you could manage to retain your place in the regt until fully recovered, but I suppose it is all for the best. It seems that your old Corps has gained a reputation second to none in the Army, and right glad am I that Burnside has received the praise due him. I would give a heap to be under him once more, or his successor Foster.

Love to all. Shall write Joe in a day or two.

Yours truly

Jule

Is your office at 29 Court?

[Note: Written vertically at the top of the back side of the letter is noted the following: JML
Jan 17
Rec Feb 4
Ans " "
1864]

Baton Rouge Jan 27th 1864

My dear John

I have now two short notes from you to answer, though one was acknowledged in a letter to Fan, I think. The other was received by me yesterday noon. I hardly think it is time to hear from that $200. So will not make any enquiries until the next mail arrives, if you have not received it by that time I will look the matter up.

I wrote Joe a letter last night giving him my views on his entering the service again, although I should prefer to have him remain at home, still I really think every young fellow should be in the Army if we ever expect to whip the rebs. There are too many shirks at the North loafing round the girls, while better men are out in the Army. I only wish a draft would be enforced that would sweep them away from their homes, then we might begin to see the end of the war. I can understand how he feels in this respect, if he should be drafted he would be obliged to go into the ranks, and sooner than run the risk of the disgrace of being a conscript, I should advise him to accept a commission in some good regiment provided he can get it. Heavy artillery is the easiest place for a fellow; light artillery to a certain extent is one of the best branches of the service, but give me cavalry for dash and excitement. It is a hard place frequently, but then one is continually howling about from one place to another, without having to foot it. I do not want to discourage anyone from entering the Army when we need so many recruits, and if the War is ever to be brought to an end, the sooner we overpower the enemy, the quicker we shall have peace.

Two years ago I expected the regts. Would all be at home in a year, but here we still remain and "the end is not yet." I made up my mind long ago to see the thing through, no matter how long I am in the Army, provided my health and strength still last, for I will know when peace is declared we shall never regret the hardships we have been through; and those cowards who won't enlist because the life is too hard in the ranks will always be sorry they did not strike one blow at the enemy. Why don't New Sumner go out again? I should think he might easily obtain a commission in some regt. How a fellow can rest quietly at home in these times is more than I can imagine. I would like to be there for a month or two to enjoy myself again with some of my old companions, but to go into business is something I could never dream of, unless I was disabled as you are.

How I wish you could all enjoy some of this glorious weather which we are now having. At noon it is almost too comfortable to be out doors; yesterday the brigade went out for a drill, but the weather being too hot we soon returned to camp. The grass is quite green in some places, and the flowers are commencing to blossom. Last Winter we had roses every day during the season, but this has been the coldest ever known in this State, so that everything is backward this year.

I expect our <u>select</u> party will have a dance this week and I am anticipating another good time. It hardly seems like war remaining at this place, out almost every evening calling on some of the girls who have changed their sentiments in regard to our troops. They have even gone so far as to give up the Southern cause, the consequence is a division of society, the Union girls, or Yankees as the sesech damsels call them and the latter. Neither will speak to the other although six months ago they were all intimate friends. It's bully for us I can assure, and a chap
time
can have a better ^flirting with the pretty girls, than sticking around camp. The people here, sesech and union all think very highly of our brigade, and sometimes we have crowds of both sexes out to see our dress parades. But I presume the time will soon come when we shall have a campaign towards Mobile or some other place, then goodbye to Baton Rouge forever.

Now that I have written you a good long letter I shall expect to receive a similar one in return, such as you used to send me last Winter. Tell Mother I will write in a few days to her; and if any of the young fellow round town ask after me, you can say to them that I am in bully spirits and down on all cowards.

Love to all. Write soon to

Your affectionate brother
Jules

Baton Rouge La Feb 27th 1864

My dear Mother

I received your letter of the 3rd last Monday and should have answered it ere this but having been down to the City, I did not feel like writing till back in camp once more. __ I am very sorry you feel so badly about Joe, But I had rather see him as he is now than have him run the chance of the next draft, when he would without doubt be conscripted and obliged to go out as a private. I know everything at home looks desolate, but how many home are there both North and South left without a single boy? All fallen in battle. I am very thankful not one has been taken away from ours, and as long as your sons are in the Army I think you ought to feel prouder of them than if they were hanging around home. I know it comes hard on my dear Mother, but remember not one of us has ever yet hung back when men were needed to strike at the rebels, and would you have Joe stay where he knows he ought not to be? The time will come, and that soon, when you will love the little chap better than ever for preferring a life of hardship to one of comfort while the War lasts. I could not remain at the North in these times, and I honor my dear bother all the more for his spirit and courage. It may seem dark to you now but the bright day will come, and after peace is declared we shall all settle down at home better satisfied for having done our duty. So cheer up dear Mother and don't feel discouraged.

I see by the papers that Amasa Guild is 1st Lieut in the 18th. He has got up quite finely. __ By the way a letter came from Fan yesterday but no papers & none have been received by me for several weeks, please tell John not to forget me. It is too bad that poor little Ned is so sick, I trust he is much better by this time; how the little fellow must have suffered. I will write to him in a few days.

Last Friday I started for N.O. and reached the City early the next morning. Saturday I went up to the Fowlers and found them all well and much changes in their feelings towards us. They were very glad to see me, more so than ever before; little Alleyne knew me in a minute and yet he hasn't seen me for nearly a year. Sunday and Monday I called again and each time enjoyed my visit very much. Monday Gilmore had the greatest concert I have ever heard in my life, over 150 pieces of the various bands were together and to crown all 40 guns were arranged so that they were fired by electricity. It was perfectly magnificent I can assure you. I had a grand time with Mr. Gilmore, and many a chat of former days in the 24th, he gave me a ticket to the great ball given on the 22nd by Mrs. Banks and to which I went in the evening. It was in th Opera House, so when tired standing near the dancers I could go up in the galleries and with my opera glass take in the whole scene at a glance. The dresses worn by the ladies were splendid though I can't say much for the fair sex myself, for I thought in such a city as New Orleans it was impossible to get a crowd of girls together without there being a set of beauties. But as the sesech fair ones were not present I suppose it is easily accounted for. Tuesday afternoon Capt Russell of our Regt., – who had his leg broken on the 14th June and has not been with us since, – and myself drove out to the Lake and took Supper. Wednesday and Thursday I passed strolling around town feasting my eyes staring at the pretty girls, of which there are crowds more than any place I have ever yet been in. Thursday evening I started for this stupid place, and yesterday noon I arrived feeling anything like settling down quietly after the pleasant week passed in the City. Fred Sampson called on me one day, but unfortunately I was out with some officers or other. Tell John if he sees him to give him my best regards and say I was sorry not to have met a townsman. By the way there is a Lieut in the 6th Cav (formerly the 31st Inft'y) who used to live in Dedham a few years ago. His name is Wm Jones, I think Joe will remember him. He goes home in a week or two recruiting maybe in D__ for a few weeks.

The campaign is commencing, the Cav. Div. with the 13th Corps, or at least the greater part of it, and the 1st Div. 19th Corps are starting from Franklin to Alexandria, while some troops & the gunboats go up the Red River. Blunt will without doubt operate with us in Arkansas, so you may expect to hear stirring new very soon. Our brigade is now changed to the 2nd (Grover's) Division; they say in NO that we will be off up the River, but we all think here that our regt will remain where it is, having only 10 officers & 250 men for duty. Our going into cavalry was stopped on order from the War Department, prohibiting all changes, so we remain "foot pads" still. Sherman is giving the rebs a great blow and our crowd will pitch in in another direction, you need not worry about our regt going, if we were they would not be detailing the Major to be Provost Marshall at this place. In spite of what is said in New Orleans about our brigade taking part in the coming Spring campaign, I have no fears of starting for we are too short of officers in the 38th to enter the field even if the brigade went. Before two weeks pass I should be in command of the regt. for Col Richardson would break down and I am next in charge. I only wish I could get back to Plaquemine for I am tired of this stupid place and do not care much whether we take the field or not, anything for an excitement.

The day I went down the river to NO I met a Capt Brulé C.S.A. who formerly went to school at Jamaica Plain, he said he had often seen me in the cars, and asked after a number of my friends there. He is well acquainted with Mrs. Weld. He told me that he had heard my Co, very highly spoken of by the people of Plaquemine & says if any of us are ever taken by the rebs around this place or below we will be treated in good shape, for they all know us by reputation. This coming from an enemy is very gratifying to me I can assure you.

I saw Maj. Schlatre for a few minutes yesterday as the boat stopped at P and he renewed his invitation. If I am ever sick or wounded I shall try to get to his plantation for it will be like home and I shall be surrounded by a crowd of friends.

It seems Mary and Reese have got hold of the absurd story, as everybody up here has it, besides all at home, ____ that I am engaged. I will tell you once for all that when I really enter into an engagement I will notify you of it at once. Just the idea of a fellow like me falling in love with every pretty and agreeable girl I meet, well it's too bad people should think so even if my letters are enthusiastic on the subject.

I sent John a paper yesterday giving a short account of the grand concert and ball last Monday, so you can tell from that what kind of affairs they were, better than I can describe them. Will write to Fan in a day or two.

Am very well, never better. Love to all & kisses to the babies. Write often to

Your affectionate Son
Julius

(Letter fragment)

Baton Rouge La March 8th 1864

My dear Mother

Yesterday was the anniversary of our first landing at this place when we came up for the advance on Port Hudson, and as a sort of celebration the 156th NY & the 38th started out for a tramp in light marching order, so that if called upon for a campaign we shall be in trim. From

two &one half

Magnolia Grove, [~~three~~] miles distant, we marched down to Craigue's Stockade (my old visiting place) without a halt, over five miles distant, and the day being hot & roads very dusty I thought my personal appearance did not come near enough to decency to permit me to call on the young ladies. After dinner we loafed around till nearly 4 o'clock & then started on the back track in a hurry making five & one half miles to our first halting place in just sixty five minutes, this just knocks any fast marching I have ever yet seen in the Army; no such time would have been made had not the two regiments been [~~marching~~] racing with each other; it used me up though but two of our officers besides myself were with the head of the column when we halted. On our return we found that the rebs had come down within sight of our pickets on the Greenwell Springs road, so this morning two of the other regts. Were ordered to go out in that direction for their exercise. They started a short distance behind the usual scouting party of thirty of the 4th Wis. Cav. Under Lt. Williams; as soon as the cavalry had got out about four miles and just at the spot where Lt. Bush was so badly wounded a few weeks ago, a large force of the rebs concealed in the cane fired on them killing poor Williams instantly and wounding several others. Finding themselves overpowered they fell back a short distance and sent in for reinforcements; when the regt reached the spot the rebs had skedaddled but were chased by the cavalry & finally overtaken then a running fight ensued & ended in several rebs being killed, some wounded and two taken prisoners. The force was scattered in every direction and driven across the Amite, but they have doubtless assembled by this time and in a few days will be round again, it's a pity we haven't Earle here now, they never were so bold before his capture.

Williams was at Plaquemine with us and as he was quartered just across the street from where I lived we saw a great deal of each other. He was a fine fellow and a good soldier; little did I think last Friday afternoon when he was in my tent to see me that we never should meet again. If his Co. had not been relieved at Plaquemine by another he might have been alive, but who could have foreseen it? The funeral will probably take place tomorrow and I shall probably attend it, for we were very good friends. He was at our Thanksgiving dinner at Plaquemine and at that time spoke of eating his next one at home, for his time was out this Summer, poor fellow little did he think it would be his last Thanksgiving on Earth.

I hope now, that an expedition will be sent to Clinton to clean out that place and destroy the town, for there is where all these gangs of rebs meet. If we only had plenty of cavalry up here we could make a big raid back of this place and break up their camps, but our present force is only large enough to do picket duty and small scouts. I'd gladly march up to Clinton again to witness the burning of the town and would like to be the one to apply the torch.

In a few days I shall send on a box addressed to John at D___ containing the following list of Articles

3 dirty undershirts	1 old shirt, the bosom & cuffs good
2 " towels	1 Pair thick drawers
1 Cape for overcoat	1 " " gloves
1 Package old letters	1 Book (gobbled)
1 Rubber pillow & case	1 Old Morning Report Book
1 Pair Wristers	1 Pair Rubber boots
1 Old Swordbelt	1 Old sack coat, the cake box, and

latter article

a bullet from Fort Butler at Donaldsonville; I wish this ^ given to Mattie Sumner as I hear she is making a collection of such things I picked it up a few weeks after the fight at Donaldsonville when the rebs stormed the works and were repulsed with such fearful loss. The spot where I found it was the scene of awful slaughter, where over 70 dead and as many wounded lay in a place a few rods square; it could tell a [?ate], and I have no doubt but that it helped lay someone out, for few shots were wasted then, as they were only a few feet apart, our men on one side the ramparts & the enemy on the other.

I thought Jennie might make a little cape for Neddie out of the one I sent home, it will do for him to play soldier in at any rate. __ The old letter please put away for me, also the boots, swordbelt, gloves (the ones Fan made me before I left in the 24th) & wristers ___ The book I gobbled somewhere in my wanderings. ___ The sack coat I want cleaned and put away; it is just as I wore it in from yesterdays march and finding the weather too hot for it now, I thought it best to get it out of the way. It is the one I wore through the whole Campaign till about the 1st of June, consequently it has seen several battles and quite a number of skirmishes, for this reason I wish it kept safely after having it cleaned. The old pillow now goes home for the third time and I hope it will stay there for it leaks so there is no use having such a thing. I also send home two towels, having more then I need, these & the shirts are dirty; the old bosom & cuffs I thought were good to put into another one. Cake box I again send back to old Dedham and hope some of these days to see it once more. ___ Tell John he will (END)

Baton Rouge March 17th 1864

My dear Mother

Your letter of the 2nd came along yesterday afternoon just as we had received marching orders, so I will write a few lines while time lasts, although as yet it is doubtful whether we get off, for other troops will have to leave [for] this place to garrison it. Gen Cooke has protested against our going so there the matter rests at present, a few more days will decide the question. I don't know where we shall go to beyond Thibodoux, for there's no necessity of marching us up through the Teche Country in the rear of the Army, may be they will put us on board transports & carry us up the Red River.

This morning a gunboat arrived bringing about 400 rebs taken at Fort DeRussey by Gen Smith of the 17th Corps; the gunboats have gone beyond Alexandria chasing the enemy's steamers and if the river rises are at Shreveport by this time. This cuts off the approach to Texas and I don't see what is to become of the reb army unless they undertake an overland march through a country badly watered. Think this State in a week or two will be cleaned out by us.

I received quite a long letter from John and will write him in a day or two. Last night the bands recruited by Gilman stopped here on their way to Port Hudson and gave us some fine music. I met two of the old band among them. Gilman has gone to the front with Banks, speculating, I expect. It is a pity that our generals would not let cotton alone and attend to their business, but everyone out here has "cotton on the brain", all they think of is making money. _____ Capt. Wyman returned here last night from a visit home; did you ever receive that small bundle sent on by him? I haven't heard any of you mention a word about it in your letters. Enclosed you will find a receipt for a box sent on to John, the freight is paid to Boston, the contents I have already mentioned in a former letter.

Am very glad to hear that Neddie is so much better, for I was a little anxious about him. I shall try to get home to see you all this Summer as soon as the campaign is over, but it is very doubtful whether I shall succeed or not, not having any influential friends to intercede for me. If Gilman is here then I can get him to "pull the wires" for me, in that case my prospects will be good, still as I said before it is decidedly uncertain.

There is nothing new going on, some say we will remain, others that we shall get off in the course of a week at the outside. It doesn't make much difference with me either way, for I am about tired of camp life and saving been here since the first of August have no right to complain of our lot. Next fall we shall probably have to take it.

Two of our officers go home to day having resigned. I almost wish I was out of the Army, but think it dishonorable to resign unless one has a good excuse, and even if I was at home I couldn't remain there long without wishing myself back in some regiment.

Well my dear Mother I must say good bye as I have a lot of work to do, requiring my immediate attention. Love to all.

Your affectionate Son
Jule

(Letter fragment)

2

In the tallest kind of a hurry else we may be all in a tight place soon. You know Stonewall Jackson always said Bank was the best commissary anyone could want, & in the Department he had kindly given the rebs everything they wanted from a gunboat down to a pistol I don't believe there is a single general on our side who has given so many stores to the enemy as Banks has, only hope he won't turn us over to their tender mercies. All the men are feeling in tip top spirits although they know we have been whipped, still in spite of this and their want of confidence in Banks & Franklin, they are willing to try once more, but whether the movement will be made up or down no one seems to know.

I saw Frank Loring last night, he is very well; passed through the whole fighting in safety.

outside the cabin

Our <u>brave</u> Col. did not show himself ^ during the few minutes the guerillas were firing at us, and the Adj hid behind some boards; it shows how well off this regt is in this respect, in time of danger they are missing, one might as well have an old horse for a commander as the person we now have.

And here my letter will have to come to an end. If we stay at this place I'll write again in a few days, but if we move forward you probably will not receive a letter for sometime. As we are now situated, I can't tell how soon we may be off.

Am in good health and only want to see the campaign successfully ended. Love to all.

Rebs acknowledge their loss was 5 to1 of ours.

In haste

<u>Jule</u>

Don't show this scrawl outside the family

Alexandria La March 28th 1864

My dear Mother

On the 19th I wrote John that we should probably start for this place in a day or two: Wednesday morning the 128th NY & our regt. with Battery C 2nd U.S. Art. Embarked on the "Laurel Hill" and at noon left Baton Rouge. Quite a crowd of young ladies came to see us off and I can assure you the thoughts of bidding all our friends goodbye were not very pleasant ones for either party. Passed by Port Hudson at 4.30 PM and at daybreak we saw the Red River; at 3. PM steamed by Fort DeRussey where the iron clads Essex & Benton are at work destroying the fortifications. About midnight tied up at this place and early the next morning disembarked, marched some three miles beyond and above the town and there pitched camp. Emory's Division arrived that day so in the afternoon I rode down to see old friends. Friday morning portions of the 13th, 16th and 17th Corps were marching by our camp on the advance to Shreveport, for seven hours a column of infantry and artillery was on the move; old regts which bore on their flags a dozen of the hardest fought battles of the war, Shiloh, Corinth, Siege of Corinth, Iuka, Jackson, Champion & Thompsons Hills, Vicksburg, New Madrid, Island No. 10, Stone River &c were prominent. It did a fellow good to see those splendid fellows marching along with that easy step marked in "veterans". That day there were at least 45.000 troops in and around this place, and Grovers division yet to arrive. I saw the eagle carried by one of the Western regts of which I have so often read, the bird they have carried in every battle. But of all the sights around here the river exceeds them by every odds; 12 large ironclads, 7 tin clads, 2 wooden gunboats and 26 transports I counted Saturday evening; yesterday they were at work getting the ironclads over the rapids and this morning the last of the troops are to leave. I presume all except the cavalry will take the transports above & go to Shreveport. Our brigade remains to garrison this post by order of Gen Banks, so we shall see no fighting at present. Saturday noon we tramped down town, and today sall have to go back a short distance holding the right of the line, only hope to get decent quarters, better than a shelter tent. Frank Loring has been to see us several times, he is looking better than I have ever before seen him.

Presume you have heard of the fight above here in which Col. Sargent 1st Mass. Cav. Was badly wounded in the leg; the troops captured 300 prisoners & 4 guns from the celebrated Valverde Battery, reckon there will be some more fighting in a few days for Dick Taylor has about 20.000 men and we double that number.

This place does not look as it did last year, the Western troops have pulled down every fence and any quantity of buildings, killed all the cattle & poultry and gobbled all the horses and mules. It looks like war. ____________ The firing heard by us at Baton Rouge about two weeks ago was at a review; little or no skirmishing has taken place during the trip through the Teche Country and but very few have been wounded.

I must now close for other business demands my attention, and then till we are settled I don't feel like writing. Am in bully health, but feel a little mad because we can't participate in the coming cotton gobble, for this campaign is but little else. Love to all.

Yours lazily
Julius

(Letter fragment)

2

every day and intend to mail one of each home and have copies struck off there for you all.

I received four papers from Fan two dated about the 6th and the other two the 12th. No letter from Joe has been received since early last week.

I can't write more now as I am obliged to go on fatigue duty in charge of building the fort. So good bye. Love to all.

In haste
Julius

Alexandria Apl 11th 1864 6 AM

My dear John

I finished my letter this morning in a great hurry, giving you the last report, supposed to be reliable, since then there have been any quantity of rumors so that this place must have been like Washington the day of the 1st Bull Run fight. It is very fortunate that there is no telegraph here from North else the whole country would be filled with news. The latest story now is that Magruder's whole Army, being driven by Steele attempted to cut their way through Banks' lines and were awfully cut up, losing heavily in killed, wounded and prisoners; also that the California Cavalry, estimated from 4 to 8000 had joined Steele. Transports filled with prisoners and wounded are now on their way from above, and that our Army is within a few miles from Shreveport. We are going to Natchitoches to form a new base and to that town all supplies will be sent, so that there's no prospect of our seeing any fighting. I am sorry that we shan't be "in at the death" for I hate this being stuck in the rear the whole time.

There is a nigger regt at this post and to show you how much better they are treated than the "white trash", I will only say they are allowed any quantity of transportation as they were at Port Hudson during that siege, while we officers have nothing but shelter tents, the privates A tents and the officers wall. Of course we all feel as if we were considered their inferiors, but its just the[y] way they always get things we can't.

As soon as supper is over I am going down town and find out the latest stories and will write you more by and by.

10 PM Since writing the above I have been howling round town trying to get the last report, but it is hard to tell what to believe as the place is filled with rumors. Otis on Grovers Staff tells me that he saw a letter from Grand Ecore dated yesterday, in which it states that our Army has suffered a temporary repulse losing a few pieces of artillery. I also learn that our Division is ordered to the front and not to Natchitoches. I am very glad of this for I dislike being in the rear; whether we shall see any fighting or not time alone can tell. You can believe what you please of this letter, for my part I think it a mixed up mess.

I received a letter from Mother, one from Joe and one from Belzy with two papers from Fan this evening. Shall write to the two former at once as we may get off in the morning. ________ You can send on my box; tell Mother I don't care for any milk as we buy it out here at a reasonable rate.

Now I shall have to say good bye for there's lots of writing for me to do to night before I turn in. Love to all.

Yours Affectionately
Jule

(Letter fragment)

2

& didn't seem to fear us at all. Rested that night & since then have been marching from 15 to 20 miles a day. Last night when I got into camp I only had four men out of fifty the rest were tired out with the hard marching & heat of the day. It is fearfully hot, I think our Summers at home hardly come up to what we are now having.

To day the Command is having a rest & I think we need it. A report is circulated that Grover has captured a lot of the rebel canon but I don't believe it. We heard that Charleston is taken I hope it is true, it is about time.

The troops are to push on for some days chasing the enemy who have scattered in every direction & numbers have been captured. You will get the news in the papers, so I shan't take up my time trying to write about what I don't know.

Love to all. Am in tip top health & decidedly tired.

Will write as often as possible.

Julius

John

Julius Lathrop grave in Dedham's Old Village Cemetery.
© Heidi Fieldston

(Above) Julius Lathrop prayerbook.
(Below) Inscription from Julius Lathrop prayerbook.

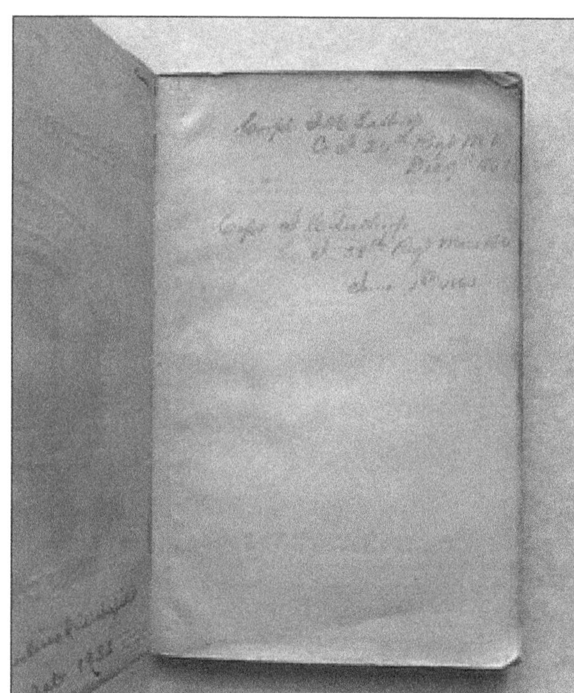

Views of Julius Lathrop Memorial window, St. Paul's Episcopal Church, Dedham, © Melanie McCarley.